PENGUIN BOOKS

SIGNS OF LIFE

Robert Pollack, a former dean of Columbia College, worked for several years with James Watson, the co-discoverer of DNA's structure, at Cold Spring Harbor Laboratory. A recent winner of a Guggenheim writing fellowship, he now divides his time between New York City and Vermont.

ROBERT POLLACK

SIGNS
OF
LIFE

THE
LANGUAGE
AND
MEANINGS
OF
DNA

PENGUIN BOOKS

PENGUIN BOOKS

Published by the Penguin Group
Penguin Books Ltd, 27 Wrights Lane, London W8 5TZ, England
Penguin Books USA Inc., 375 Hudson Street, New York, New York 10014, USA
Penguin Books Australia Ltd, Ringwood, Victoria, Australia
Penguin Books Canada Ltd, 10 Alcorn Avenue, Toronto, Ontario, Canada M4V 3B2
Penguin Books (NZ) Ltd, 182–190 Wairau Road, Auckland 10, New Zealand

Penguin Books Ltd, Registered Offices: Harmondsworth, Middlesex, England

First published in the USA by Houghton Mifflin Company 1994
First published in Great Britain by Viking 1994
Published in Penguin Books 1995
1 3 5 7 9 10 8 6 4 2

Copyright © Robert Pollack, 1994

All rights reserved

The moral right of the author has been asserted

Printed in England by Clays Ltd, St Ives plc

Except in the United States of America, this book is sold subject
to the condition that it shall not, by way of trade or otherwise, be lent,
re-sold, hired out, or otherwise circulated without the publisher's
prior consent in any form of binding or cover other than that in
which it is published and without a similar condition including this
condition being imposed on the subsequent purchaser

CONTENTS

ACKNOWLEDGMENTS

At the heart of this book is a gift from my wife, Amy, and our daughter, Marya: the time, space, and freedom to write, and the encouragement to begin again.

It was an immense privilege to work with John Sterling at Houghton Mifflin, whose insight and talents have surely made this a better book. I am indebted as well to my agents, Anne Engel and Jean Naggar, and to our mutual friends, Penny and Horace Judson, without whose kind and thoughtful help little would have been possible.

I thank the Alfred Sloan Foundation, the Howard Hughes Medical Institute, the Abe Wouk Foundation, the John Simon Guggenheim Foundation, and the Dartmouth College Department of Anthropology for their support, and Columbia Science Librarian Kathleen Kehoe for engaging the metaphor of this book so congenially by eagerly getting me every book and paper I ever needed.

David Albert, Akeel Bilgrami, Suzie Chen, Eric Holtzman, Nancy Hopkins, Ken Korey, Joshua Lederberg, Madeline Lee, Michelle Mattson, Alan McGowan, Don Melnick, Hugh Nissenson, Bruce Pipes, Michael Rosenthal, Charles Sheer, Ted Tayler, Jim Watson, Amy, Marya, and my brother Barry were all kind enough to make many useful suggestions; all the oversights, wrong guesses, and misunderstandings that slipped past me are my own responsibility.

INTRODUCTION

THIS BOOK BEGAN in the late 1970s, when my family and I first learned that my father was afflicted with Alzheimer's disease. He has since died, after a long period during which he hadn't enough of a mind to recognize us, a source of much despair for my family and me as we pondered the mystery of this slow, sad end to his life. When he was diagnosed with the disease, I did what any sensible molecular biologist would do. Trained to read the literature of science and medicine for clues on how to attack such a problem, I spent much time in the library of Columbia's College of Physicians and Surgeons. Not realizing how unhappy this was making me, I persisted in the notion that I could help my father if only I could find the pattern that I knew had to be hidden in the published research literature. I was wrong; there was no discernible path to reversing the course of this disease ten years ago, and while there have been some recent tantalizing flashes in the fog, there is no path I can see today. This was hard for me to believe then, and it still is. But believing it, and accepting it, was necessary to having any sort of a life for myself. The life I have had since his disease was diagnosed is quite similar to the one I had before, with one clear difference: I still believe we can figure out how the natural world works through science, but its limits have become as interesting to me as its powers, as I realize that scientific discoveries cannot protect me from the pain of loss.

I have always loved books: I have been a scientist since high school; before then I read omnivorously and wanted to be a writer. My grandparents had been driven from Eastern Europe at the turn of the century, arriving at Ellis Island with little besides hope for an auspicious future in America for their children if not for themselves. Like so many other children of immigrants who fell short of their parents' dreams by leaving school to work, my parents held tightly to the nineteenth-century notion of science as the motor of progress and the leveler of hierarchies. The authors they read and passed on to me — H. G. Wells, Paul de Kruif, J. B. S. Haldane — led me to believe for a short while that scientific laws had been found to govern all history and all behavior and these laws were as predictable and general as those that governed energy and matter. My adolescent rebellions had a peculiar twist: as my friends broke the dietary rules of their parents, tremulously eating their first pork in a Chinese restaurant, I staked out my independence by insisting that scientific laws did not extend to human affairs and that history was not predictable. From this mutiny I formed the idea of becoming a scientist myself, a radical notion for a boy whose pleasure came from reading and whose parents had not finished high school.

I was just preparing for high school myself in the spring of 1953 when James Watson and Francis Crick published a nine-hundred-word prose poem in the British journal Nature that established the symmetrical, twisted structure of the chemical deoxyribonucleic acid — DNA — and showed that the way it was made explained how it could serve as the chemical of the gene. Of course, this purest and most beautiful example of the inextricable intertwining of form and function in biology made no quick headway in the educational establishment responsible for my schooling, so biology remained for me a confused subject full of lists and mysteries. At Columbia College I chose physics as my science, because it seemed to me to be of the greatest generality and therefore most likely to yield the underlying laws of nature I had decided to pursue. But by the time I graduated, the explosion of new ideas and results in molecular biology and genetics had reached even me. I turned to biology as a graduate student early enough to learn about genes and their chemistry from research papers and seminars given by their authors, not from textbooks. From physics I brought two contradictory lessons: to guard against unnecessary complexity and to be constantly aware that what was easily observable was no guide to what was invisibly small.

I began my career as a biologist studying the genetics of bacteria and the viruses that live within them. But as soon as I finished the project in bacterial genetics that earned me a Ph.D. from Brandeis University, I found myself drawn ineluctably to the border where science touches directly on matters of human concern. I joined the Pathology Department at New York University School of Medicine to begin what would be a thirty-year project: to unravel the molecular events that led to the appearance of a cancer in an otherwise healthy tissue.

In the late 1960s, after two years of work, I made my first personal contribution to the problem in a paper showing that within a tumor there would always be more normal cells, and that these spontaneous revertant cells had regained the growth controls that kept a normal tissue from overgrowing its boundaries. This paper caught the eye of James Watson. After one seminar and two beers, he invited me to join a small group of young scientists at Cold Spring Harbor Laboratory, a private research institute on the North Shore of Long Island that had, as I remember, two Nobel laureates on its permanent staff of twelve when I arrived.

In this way I enlisted in President Nixon's War on Cancer, a baroque moment in the history of American science that provided hundreds of young scientists with jump starts to our careers. When I ran a lab at Cold Spring Harbor in the early 1970s, my family and I lived in a set of converted stables. There were weeks when I would walk from our apartment down the laboratory's main road to my dishes of normal and cancer cells, then back up the road at night to our vegetable garden, not knowing which day of the week it was, not carrying any money or keys (let alone any ID or credit cards), just drunk on science all the time.

I would sometimes meet Barbara McClintock — perhaps the most original geneticist since Mendel and certainly the most self-disciplined person I have ever known — on my walk to the lab if I were up early enough, and she would show me her flowers. She was interested in the genetics underlying the appearance of the little red flowerlets at the center of a bunch of Queen Anne's lace. She would offhandedly teach me about those flowers along the road in ways that quietly conveyed to me that she knew each of them, personally and by name, and that she also knew every other plant within sight and its family history as well.

At home in this scientist's Arcadia, I continued my work on simple versions of cancer, imitations of the disease in a dish. I grew cells from mice and people, from normal tissue and tumors, seeing to it that each cell preserved the characteristics reminiscent of its status in the body. I took DNA from various sources — a cell, a virus, a bacterium — put it into another cell, and followed the inherited changes. Over the years, this premeditated genetic transformation brought on by moving DNA from one cell to another permitted my colleagues and me to describe in daunting detail the genes that change when a normal cell of the body turns into a cancer cell.

In the 1970s and later, as my father was dying his slow death, scientists in my field added to DNA transformation a second skill of comprehensive importance when they learned how to change a gene as they desired. Since then, the direct manipulation of inheritance — an impossibility when I was a graduate student — has become the stuff of high school laboratories. Molecular biologists, including younger colleagues who were not even born when I was in high school, now hold great power over inheritance. Some have begun to transliterate the entire sequence of human DNA — the human genome — into a string of English letters; others have succeeded in recasting the DNA of other species of plants and animals, including many of the ones we eat.

I'd bet that if we survive long enough to be able to look back, the second half of this century will be known, not as the Age of Atoms, but as the Age of DNA. Just forty years ago Watson and Crick were making tin and paper models of DNA and uncovering the set of laws that govern all inheritance, the base-pairing rules that make DNA the perfect self-copying text. Watson's life has encompassed the Age of DNA: a few years ago, he brought together the people and the money necessary to launch the comprehensive transliteration of human DNA, the Human Genome Project. Already, the new tools of molecular biology are capable of making profound changes in the human genome, of changing us and giving us the capacity to change our children for better or worse. This new capacity has come upon us very quickly: it is as if we had just deciphered a few words in a new language and begun rewriting ancient texts before understanding their full meaning. The task that lies beyond the Human Genome Project is not merely to transliterate, nor simply to comprehend a cell's way of reading its DNA, but also to translate, and then to read, the DNA within ourselves with the full analytical power we might bring to any of our own texts.

My science — molecular biology — has become to a large extent a project to understand fully how our bodies read the unique and uniquely interesting chemical texts within each of our cells. The meanings we have begun to draw from human DNA and the changes we have begun to make in it are important and should be understandable to everyone, which is one reason I have organized this book around the notion of human DNA as a work of literature, a great historical text. But the metaphor of a chemical text is more than a vision: DNA is a long, skinny assembly of atoms similar in function, if not form, to the letters of a book, strung out in one long line. The cells of our bodies do extract a multiplicity of meanings from the DNA text inside them, and we have indeed begun to read a cell's DNA in ways even more subtle than a cell can do.

The reality of "deep time" — that is, that the history of the planet goes back tens of thousands of millions of years and that life has been on the planet for thousands of millions of years — is at first difficult to comprehend. Looking at life in all its variety and complexity, the alternative is to believe in magic. The notion of DNA as a text makes it possible to imagine natural selection as an author in deep time, writing at the rate of perhaps a letter every few centuries to produce the instruction books for all the living things we are among today. Just as an analysis of the fossils in rocks that now show themselves on the surfaces of the planet gives us the actual historical record of mountains and oceans as they formed and reformed over the last couple of billion years, so an analysis of DNA texts will provide direct information about events in biological evolution that shared the same deep times. Before long, museums of natural history will be in practice what they have always been in theory: libraries of DNA, with holdings that range from current editions to the ancient and unexpected DNAs of fossils.

Our own species is young only in comparison with life itself; we are all the children of ancestral peoples who walked the earth hundreds of thousands — perhaps even millions — of years ago. Each living person's DNA is rich in specific passages derived from a particular genealogy. Yet at the same time we can be sure that the texts we find will all refer to the same past, a past of branching descent. We must begin to see the texts of an individual and the texts common to members of a species as a form of literature, to approach them as one would approach a library of precious, deep, important books. A proper appreciation of molecular biology as one of the arts — a branch of history — as well as one of the sciences may provide the burgeoning field of molecular medicine with the manifold vision of the world that both science and medicine need if they are to honor the great Hippocratic constraint, to do no harm.

～

When a colleague in Columbia's Philosophy Department asked me what this book was about, I said that it was an attempt to explore the implications of the fact that we are on the verge of grasping at least some of the multiple meanings of our DNA. He responded, "Must I find out?" We both laughed, but afterward I could not shake off the troubling sense that he was casually conveying a deeply felt but usually silent doubt that many sophisticated people have about the utility and necessity of understanding even the science most likely to affect their lives. In a 1993 national Louis Harris poll sponsored by the March of Dimes, 86 percent said that they know "almost nothing" or "relatively little" about gene therapy. Scientists too have often remained silent, by and large failing to make clear to a larger audience what it is they do, why they do it, what they expect to accomplish, and what they fear. Why has it been so hard for scientists and nonscientists to communicate productively? There is no simple answer, but in writing about the meanings of our genes I have found it necessary to push against three barriers to mutual comprehension, even with the metaphor of DNA as text to ease the problem of translating the jargon of my field.

First, many nonscientists are wary of scientific language, often recoiling from its authority and sometimes fearing that complicated notions are being used to legitimize unspoken political agendas. Without an understanding of the underlying science, wariness turns to an unfocused distrust. In the same Harris poll, 73 percent agreed "strongly" or "somewhat" with the statement that "the potential danger from genetically altered cells is so great that strict regulations are necessary"; the poll did not ask about the internal inconsistency of the majority's response, which coupled an admission of ignorance about genetics with a certainty of its riskiness. But wariness has been called for in the past, and if science is to be kept honest, there is no harm in approaching it with a well-informed skepticism.

Consider, for example, the damage done by geneticists in the period leading to the U.S. immigration laws of the 1920s. These laws were "scientific" since they drew their rationale from the sworn congressional testimony of many highly regarded biologists and physicians. Committees of Congress heard "scientific" proof that similarities of language, religion, or ethnicity were inherited expressions of a small number of underlying "genes," that such undesirable traits as laziness, drunkenness, avariciousness, and poverty were simple genetic traits. In this way, scientists gave intellectual cover to the decision to exclude immigrants on the basis of their skin color, religion, nationality, and language: it was Italians who were genetically more likely to be poor, Jews from Eastern Europe who were genetically greedy, and so forth. The inaccuracy, intellectual sloppiness, and prejudices of these scientists and like-minded members of Congress converged in the Immigration Law of 1926, which codified the most crudely racist and biologically foolish distinctions since the Constitution's definition of a slave as 60 percent of a human being. By the

1940s, this eugenically correct law had blocked the escape to the United States of many people who subsequently died in actions carried out according to the more activist laws of the German Third Reich.

This impulse to misuse the language of genetics for political purposes has not disappeared: the "ethnic cleansing" of Bosnia is built on the same misguided dream of biological purity. Nor has the United States freed itself of the problem. In 1992, federally sponsored meetings were called to discuss the likelihood of discovering "genes for violent behavior" and the ways in which such genetic information — not that it is likely to exist — might be used to finger "susceptible" people and their families even before any crime had been committed. The resulting publicity led the granting agency — the National Institutes of Health — to cancel one of these meetings, but we have no assurance that it was the last of its kind.

Practicing science is an analytical act and also a creative one. The second barrier to understanding is that scientists' endless rounds of "on the other hand" are often mistaken for an inability to be certain about the meaning of an experiment. It is as though every professor of English were at once a writer of novels and an interpreter-theorist of those very novels. Such people are rare in the humanities and social sciences, but all productive scientists are quite resigned to oscillating between creation and criticism, knowing that public recognition without continued carping from one's peers is useless and empty praise.

The basic problem is that most nonscientists do not have a very clear understanding of the paradox underlying all scientific advances: namely, that scientists love to do experiments that show their colleagues to be wrong. By this adversarial process, science gradually reveals the way nature works. The notion that published science must be free of error, and that error itself indicates sloppy thinking or fraudulent intent, is misguided. Bystanders often misunderstand the place of error in science and imagine that scientists who override one another's findings are in some way not entirely serious about their work or that no scientific statement can be true if any one is false.

The willful disregard by society at large of what has been

discovered by science is the third barrier to serious discourse. When political representatives ignore the facts they pay scientists to discover, those scientists who wish to help society benefit from their discoveries feel duped. Take the matter of prenatal care: bad or nonexistent prenatal care can cause a lifetime of misery, often accompanied by a huge societal bill. Scientists have explained this perfectly well at the cellular level: the final organization and function of the cells of the body are at greatest risk when tissues are being formed, and such cellular organization continues into the period immediately after birth. Prenatal care begins before a woman becomes pregnant: to assure that her newborn child will have the full use of its endowment of intelligence and good health, a woman must eat proper food and abstain from drugs and alcohol. Yet our country still refuses to acknowledge a nonjudgmental duty to educate every woman about her body and to nurture every pregnant woman and every newborn child.

Or take the issue of biological diversity. The 1973 Endangered Species Act is based on the notion — endorsed by most members of the scientific community and buttressed by new information at the level of DNA — that the loss of a species may affect all other living things, including people, because species evolve and survive through mutual interactions and competition. During George Bush's term as president, Manuel Lujan, Jr., the secretary of the interior, was in charge of enforcing the act. In May of 1992 the *New York Times* reported that "Secretary Lujan in fact does not believe in Darwin's theory of evolution, with the rise and extinction of species." We can only wonder whether the secretary — and through him the Department of the Interior — saw the Endangered Species Act as sensible or whether to him it was fatally flawed by virtue of its grounding in the fact of evolution. In the face of such official attitudes, one can easily understand why many of my colleagues have little incentive to try to explain their work to the world at large.

～

In the early 1970s, when every transformation of a cell by an altered DNA was a novel event, I found myself in a situation that altogether changed my way of seeing science as a calling and as a profession. One of the first genomes to be spliced into a recombinant DNA molecule belonged to the tumor-causing virus SV40, which was joined to the genome of a plasmid of the bacterium *E. coli* in Paul Berg's laboratory at Stanford in 1971. This new genome combined genetic information from a bacterium with genetic information from a virus that normally lives in the kidneys of African monkeys. Such information had never been found in one genome before; the closest version of it in nature had not existed since the age when the only living things were neither bacteria nor animal cells nor their viruses, but some precursor of them all.

News of this accomplishment reached me at Cold Spring Harbor, where I was teaching Janet Mertz, then a graduate student of Paul Berg's, now a professor at the University of Wisconsin. I was also studying SV40 and thought it possibly a dangerous virus; closely related viruses, called JC and BK, had been found at autopsy in the brains of many people. It seemed to me that to put SV40 genes inside a laboratory culture of one of the bacterial species that colonize our intestines risked accidentally transforming someone's colon cells by SV40 DNA. This would be a new route for these genes, one our bodies were not prepared to defend against. Concerned, I called Professor Berg and asked him whether he had thought about the possibility of these risks. His first reaction was one of controlled astonishment at my sheer effrontery, but he did listen. After a few more phone conversations, he agreed to suspend further experiments, and to recommend that others do the same, until the recombinant plasmid could be tested for safety.

Berg's statesmanlike willingness to pull back voluntarily from an exciting line of work on the basis of a totally hypothetical and certainly low risk was unprecedented and significant. It led to a program of tests carried out for the NIH in military facilities like Fort Detrick in Maryland, where workers were protected by vents and shields that had been developed to insulate technicians engaged in producing germ warfare agents. The very air leaving these facilities was passed through a flame to prevent the escape of any microorganisms. After more than a year the results were clear: the many recombinant plasmids, viruses, and bacteria tested were each no more — and sometimes were less — infectious than the most infectious of their original sequences. More to the point of my initial concern, the intestines of volunteers who ingested laboratory strains of recombinant *E. coli* did not, in fact, become overgrown with these bacteria; the normal bacteria of the gut prevailed. Once the test results were in, the NIH decided to allow recombinant DNA research to go forward, but it established a Recombinant DNA Advisory Committee to serve as a watchdog and clearinghouse for new developments. In the past fifteen years the group has not had much to do, and its purview has been slowly but steadily reduced; whether this is wise remains to be seen.

Professor Berg was able to continue his work unimpeded by further phone calls until 1980, when Stockholm called to say he had won the Nobel Prize in Chemistry for his discoveries in recombinant DNA. But from the time I called him to the present day, I have never been able to feel entirely comfortable with one of the basic premises of science as it is currently practiced. The concept of peer review is built on the notion that scientists alone should judge one another's work, but that phone call to Professor Berg was just too hard for me to make. And though I am sure I had every right to query him as I did, I have often wondered whether I would have called if I had been competing directly with him at the time. I think not; after all, it was Berg, not I, who agreed even temporarily to stop a most interesting line of work.

~

Back to the philosophy professor's question: Why poke around inside myself to get at a biological text, magically small though it might be? Are we so driven by hubris? Why not leave the text alone, unread? Curiosity is one reason, but humility is a better one, and awe is better still. Evolution has taken about four billion years to write the set of texts we can find in the DNA of creatures alive today. We are about to be able to read them in much the same manner we read any book of our own creating. This is a daunting prospect, but I cannot imagine any people so devoid of curiosity about their own bodies that they are not interested in what the text says. Nor is there really any choice in the matter: once we begin to read the book that describes how we ourselves are made, it is unlikely we will stop in the middle. The real question is not whether, but how, to read the human genome.

In *Signs of Life* I have tried to show how biology and medicine have come together in the project to understand how our bodies read DNA and how this new biology may be superseded by yet another perspective, as boundaries to the meanings we have begun to draw from DNA and the editorial changes we have begun to make in it become apparent. I have also considered the implications of one unexpected consequence of the new biology: its intersection with another, completely separate intellectual movement that has grown and flourished in the last few decades. With the discovery that a set of symbols has been used by nature to encode the information for the construction and maintenance of all living things, semiotics — the analysis of languages and texts as sets of signs and symbols — has become relevant to molecular biology. Semiotics has given students of the DNA text a new eye for reading, allowing us to argue for the validity of a multiplicity of meanings, or even for the absence of any meaning, in a stretch of the human genome.

Although semiotics and molecular biology both have been remarkably fertile in recent years, few scientists or literary critics have been prepared to move out of their own familiar territories to learn from the other. As a result, each kind of text has usually been analyzed by someone trapped in one or the other set of unnecessary intellectual constraints: the critic does not know that nature too has invented ways to read meanings into a text, and the scientist does not fully take the point that the transliteration of a DNA sequence into a string of four letters is no more likely to reveal the multiplicity of meanings in a gene than the transliteration of a poem by Pushkin from Cyrillic to English letters would enable an English-speaking person to see the layers of meaning in the poem. Yet if each strongly believes one type of text is worth reading, it is because both types of text — literary and genetic — may touch on the same matters of consciousness and mortality.

Despite my often-recollected helplessness in the face of my

father's illness and death, I have tried to keep molecular biology in perspective. I have never seen science as a religion or merely as a business or a game. Nor have I ever been tempted to join my colleagues from the other side of campus who dismiss science as "nothing important — life is not an experiment." Now past fifty, I am as certain as I was thirty years ago that this branch of science will continue to yield answers to ever more important questions. Once DNAs are read like books, whole new worlds of discourse will open up as we borrow from various disciplines involving textual analysis to interpret nature's text. How does a cell read its DNA? Certainly with great precision and with greater fastidiousness than we can read and agree on the meaning of our books. Are cells reading, exactly, or are they merely decoding? If DNA encodes our ability to read, how can we read that information itself? Will there be, for instance, a "canon" of DNA texts? Will DNA texts allow themselves to be analyzed for their meaning, or will meaning be evanescent and in the mind of the reader, as audience theorists today would predict? When we can read DNA the way we read a book, we will have to address these questions.

Though the metaphor of DNA as text opens many doors, the walls of ignorance and indifference between science and the rest of our culture make it hard for anyone to feel comfortable with the complex task at hand. Humanists shirk a text constructed by natural selection and written in an invisible chemical medium. Scientists avoid projects that cannot be framed as questions to be answered by controlled experiments. Still, the effort will have to be made. We and other living things are, after all, united by a common past and a common chemistry: we are related enough to a duck and an orange that we can eat them both. Now that we can make an original text by hooking together DNA from ducks, DNA from oranges, and, if we wish, DNA from people, it is time for everyone to learn about the language these DNAs share, the dialects they have evolved, and the arguments they articulate.

As the chance to determine our origins and our history becomes a necessity, we will all have to change the way we understand the essential nature of our bodies and our minds.

In my own effort to do so, I have taken inspiration from the English poet Robert Graves, who once wrote to me:

> Poets and physicians are closely allied in thought. Diagnostics and cure (truth and love, in essence) belong to both professions. In fact I find real doctors far closer to me in spirit than musicians or painters or sculptors. By "doctor," of course, I include all scientists who are not routineers of science, but have hearts and minds and are finding out the relations between mind and its physical concomitants.

1

INVISIBLE CITIES AND

CRYSTALLINE BOOKS

MANHATTAN IS A ROCKY ISLAND at the mouth of the Hudson River. Since 1810, when it was cut up into convenient rectangular blocks for future sale, it has accumulated large buildings. The two largest — not the most beautiful, but the largest — are located on the Hudson shore at the southern tip of the island. There, on their own windblown, rather desolate plaza, stand the twin towers of the World Trade Center. They are huge. Close up, they loom over a visitor and seem to be toppling from their own vast bulk. Inside, they are like most other office buildings of the late twentieth century: anonymous, filled with endless offices, each in turn brightly lit, well vented, but depressingly the same from floor to floor for more than a hundred stories. Each floor of each building covers almost an acre, which is cut up into a three-dimensional rectilinear grid pierced by dozens of elevator shafts. The outer wrapping of each building is made of metal and glass and is permanently sealed against the elements except at the base, where pipes bring in and take out the necessary fluids, energy, and heat, and at the top, where the buildings vent their excess heat. Prodigious numbers of people and vast amounts of information and money move in and out of these buildings every day, but at the end of each day they remain, quite unperturbed, much as they were on the day they opened.

Now imagine for a moment that we have slipped into a

slightly altered version of reality: we visit the World Trade Center plaza and find only one tower before us. Confused, because we recall there ought to be two towers, we go inside. We find the expected warrens, corridors, elevators, and such, but no sign of another tower. In our wandering, we come upon an unexpected set of barriers. We have entered the central core of the building, which is set off from the rest of the floors by an elaborate array of restraints. Authorized personnel scurry back and forth through sets of double doors, carrying sheaves of computer printouts. It is not at all clear what work they are performing.

Sneaking past the barriers, we find ourselves in an odd sort of library. We wander around the stacks, noticing that there are almost no books. In fact, the stacks themselves hold only duplicate sets of a series of books, two copies of each volume of what looks very much like the 1969 edition of the *Encyclopaedia Britannica*. Officious staff are carefully taking down one or another volume, photocopying an article, and replacing the volume. These clerks come and go, carrying out photocopies of pages of the library's twenty-three-volume encyclopedia and coming back with instructions for more copies, perhaps of the same article, perhaps of another. It all looks very orderly but also peculiar. It seems odd that the library should be sealed off, and odd also that so many copies of so few articles are being made.

Exploring further, we come to a separate sector of the library, where volumes of the encyclopedia are being painstakingly reproduced, bindings and all. This work is nearly complete: as we watch, the final volume in this duplicate set of volumes is finished, and all forty-six books are bundled off to a far corner of the library. We try to follow, but a sealed door stops us. Behind us, the library staff is suddenly excited; some people are scurrying out of the library, and we slip out in their wake.

We walk toward the elevators, puzzled by a certain soft vibration underfoot. The elevators are all temporarily out of order, so we enter an office to look out the window. We are just beginning to count ferries in the harbor when, quite smoothly and silently, the entire tower splits down the middle from top to bottom. There, just out the window and neatly aligned with ours, is a second tower on its own place on the plaza. No one in the building but us seems to find anything unusual about this quiet, massive, precise doubling of a sky-scraper. The elevators start up again; we return home to take a troubled nap, dreaming of the moment when the two towers will each divide again, cramming four onto the plaza overlooking New York Harbor.

~

Living objects all share four attributes: they can produce off-spring, they have a history of common descent from shared ancestors, they are made of invisible soft building blocks called cells, and each cell carries within itself a singular chemical called deoxyribose nucleic acid, or DNA, a large molecule assembled from the atoms of just five elements — carbon, phosphorus, nitrogen, hydrogen, and oxygen. The eye has its limits: without help it cannot see the parts — the atoms linked to one another as molecules, the molecules gathered into cells — of which it and all other living tissue is built. As a result, it has taken us a long time to see that the first three attributes are consequences of the fourth: of all the molecules in every cell, DNA alone unites all life in a common history, because every cell of every living thing has for about the past four billion years contained a version of DNA.

In human beings, as in every other living thing, DNA tells each cell exactly how to produce the thousands of other molecules that maintain the cell's shape and its place in the body. It is DNA that tells a liver cell to stock the blood with fresh proteins and to store sugar, a nerve cell to stretch itself out into long threads whose tips communicate with the others they touch. And as the metaphor of the replicating towers suggests, the DNA of a cell contains the instructions for making precise copies of itself and of the cell it is in. Each of us has always had, in each of our cells, a DNA text that guided our development from fertilized egg cell to embryo, fetus, and person; it is a precise copy of our sole and complete inheritance, one that is far more ancient than any human artifact. Because of its remarkable ability to command its own precise replication, DNA has sustained and linked all people since our beginnings as a single species at some time over the past million years.

DNA governs the operation of every cell in our bodies. A cell is a busy place, a city of large and small molecules all constructed according to information encoded in DNA. The metaphor of a city may seem even more farfetched than that of a skyscraper for an invisibly small cell until you consider that a cell has room for more than a hundred million million atoms; that is plenty of space for millions of different molecules, since even the largest molecules in a cell are made of only a few hundred million atoms. DNA ensures that a cell is not just a chemical soup but a molecular city with a center from which critical information flows, a molecular version of King David's Jerusalem. That walled city, with its supply of food and water entering through special portals and channels, had a great temple at the center and a book at the very center of the temple. A cell's version of the temple is the nucleus, a membrane-wrapped receptacle enclosing the cell's DNA. The nucleus is also the hub from which portions of the text are delivered to the cell, just as the sacred scripts were read to the people of Jerusalem from the entrances of the temple.

Observed through a powerful microscope a cell looks very complicated, and it is. It is wrapped in a fatty outer membrane pocked and studded with molecules — gates, channels, receptors and probes — that enable cells to be in contact with one another and with their immediate surroundings. The territory between the nucleus and the cell's outer membrane is called the cytoplasm — a maze of rods, balls, and sheets studded and filled with enzymes and various other proteins, all suspended in a salty gel.

Early analysts of the body's chemical composition did not place much importance on DNA, since it makes up such a small fraction of the material of the cell compared to its salts, proteins, carbohydrates, and fats. If a cell were as big as the Old City of Jerusalem, each chemical "letter" in the cell's DNA text — consisting of a few hundred atoms — would be about as big as a letter in a word of any familiar book. Yet every part of the cell, no matter how complex its form or function, is made according to information contained in the DNA folded into its nucleus. To appreciate this triumph of molecular origami, consider that the DNA in one human cell, if unwound and straightened out, would be a pair of molecules each about one yard long. A yard of DNA is a hundred million times longer than it is wide, and this exquisite thinness is the key to its ability to fit inside the nucleus. The pair of yard-long DNAs in a human cell are so slender that about ten billion copies, laid side by side like the wires of a telephone cable, would fit inside a waist-length human hair. That is about as many pairs of human DNA as there are people on the planet today; a genetic archive of our entire species could therefore be tightly packed into one long human hair if we had the means (and the desire) to do it.*

The DNA in every cell of a person — called the person's genome — is close to an encyclopedia in design and content. Like any proper encyclopedia, a human genome is divided and subdivided into volumes, articles, sentences, and words. And as in an encyclopedia written in English or Hebrew — but not a logographic language such as Chinese — words are further divided into letters. Biologists call the volumes in a genome chromosomes, a word derived from the Greek for "colored bodies," because they were large enough to be seen as dark dots by the first people to look at cells under a microscope. The articles in a genomic DNA text are the sets of genes that interact to give a cell or a tissue its specific character, and the sentences are the genes themselves. The words are called do-

---

* We are very big compared to a cell, and cells are very big compared to the atoms of which they, and we, are composed. We are made of about 100 million million, or 10^{14} cells, and each cell is made up of about 10^{14} atoms. Put another way, the complexity of a cell in molecular terms is about as great as the complexity of a person — brain and all — in cellular terms. This is as hopelessly unintuitive as the fact that the universe is hundreds of millions of times older — and life on earth is tens of millions of times older — than the oldest living person. The universe is about 3×10^{10} years old. The planet Earth has spun around the sun for about 4.5×10^9 years; life has left remnants of its existence that are at least 3.8×10^9 years old, telling us that Earth spent relatively little time as a lifeless planet. The billions of years that have gone by since Earth first held life make up the "deep time" of joint geological change and biological evolution, against which all human time, measured in millennia, is as leaves of grass.

mains, and the letters, base pairs. The concatenation of letters
into words, words into grammatical sentences, and sentences
into articles occurs in DNA, but in order to see how, we will
have to first know how the letters themselves are formed and
how they can make good copies of themselves.

∾

The 1969 edition of the *Encyclopaedia Britannica*, which
played the role of the human genome in the story of the
replicating towers, has twenty-three alphabetically ordered
volumes of articles that altogether contain about two hundred
million letters. Most of our cells have pairs of each chromosome; the twenty-three pairs contain about six billion base
pairs, so a single human genome is a text about three billion
letters long. In each volume of an encyclopedia the string of
letters is organized into thousands of separate articles about
discrete subjects. In the long string of DNA letters in a chromosome, there are thousands of stretches of letters — genes —
that each address a particular topic: how to make a particular
protein perhaps, or how to find another stretch of DNA. The
index of the *Britannica* has about two hundred thousand entries. Altogether, the chromosomes of a person contain at least
a hundred thousand genes; we do not yet know exactly how
many there are.

The topics in an encyclopedia are ordered by their spelling,
rather than their meaning, so that a reader can quickly find the
right place without knowing much about any given topic.
Sometimes, especially when the topic words themselves have
a common origin, adjacent topics may be related by meaning.
Similarly, the genes in each chromosome are present in a precise order that seems usually — but not always — to be arbitrary; when genes that do similar things are next to each other,
they are likely to be related by common descent from a single
gene. The orderliness of genes in each chromosome shows up
during the division of one cell into two, when the DNA of each
chromosome coils up on itself, giving each one a characteristic
set of crosswise bands. Each band marks the presence of a few
hundred genes; the pattern of bands on each chromosome is
very regular from person to person.

The constancy of chromosome banding from one to another individual within a species establishes that the genes are in a specific order, but it also masks a great deal of individuality in each person's DNA. Because the genes themselves are too small to see with a regular microscope, their textual differences from person to person are not normally visible within the bands, just as the differences among people on the ground are not visible from the top of a tall building. Rarely, a person is born with one or more chromosome bands out of place, snipped from one chromosome and attached to another, or duplicated as an extra copy within the genome. Such errors are like blocks of articles accidentally bound out of alphabetical order in an encyclopedia volume. Perhaps because misplaced genes cannot be found by the body in much the same way users cannot easily find an article if it is not in its proper — though arbitrary — alphabetical place, such anomalies are almost always accompanied by abnormalities at birth.

The notion of DNA as a text is far more than a metaphor. The letters of a human genome do encode more information than the *Britannica*, and both genome and encyclopedia carry their information in a single string of letters. We don't think of the book's letters as a single string because they are separated into words, sentences, lines, and pages. While the paper, the size and font of the type, the length and number of lines, the amount of white space between the letters and lines of type, and the binding all contribute to the pleasure of holding and reading a book, they serve first and foremost to preserve the correct sequence of letters and spaces from beginning to end.

DNA and the letters of a book are alike in function though not in form: the meaning in a DNA molecule emerges only when its genes are read in a useful — not necessarily alphabetical — order. The linear order of genes in a chromosome band need not be the order in which they are read out in any cell; genes, like sentences, are separate and separable passages, each critical to the way different cells derive different meanings from the same long, complicated text. In each chromosome, line and page breaks do not occur; the letters form one continuous line. But along that line, the genes are separated by other sequences that serve as molecular punctuation marks. Sets of genes form arguments the way sentences form paragraphs and articles.

Magnified a million times, the letters in DNA would be about the same size as the letters in this sentence, but the text would not look the same at all. The linear sequence of letters on a page is two-dimensional. We read with our eyes, seeing and understanding sets of letters and spaces as words. Then we use a mental lexicon and the syntax of our language to understand: first we grasp the words and the grammatical sentences they form, then the paragraphs and chapters, and then, finally, the arguments of the entire book. DNA too has words, syntax, and meanings, but as a text it is a molecular LEGO set, a sculpture whose information — encoded in its very shape — cannot be read at a distance. It has to be "felt" by other molecules in the cell in order to be read. DNA's letters are three-dimensional; they are read and understood by touch, the way a blind person reads a Braille text.

Now imagine DNA enlarged a hundred times more — a hundred million times altogether — so that its letters are as big as pizza pans. At this magnification, the human genome would have about the same circumference as a person's waist: it would be about fifty thousand miles long, and its atoms would be the size of marbles or golf balls. Its surface would be knobby, not at all smooth, and you would immediately notice two twisted ridges running the length of the DNA, like two vines wrapped around a tree trunk. Looking closer, you would see that the twisted ridges — each as thick as your arms — were a repeating chain of carbon, hydrogen, oxygen, and phosphorus atoms. When you felt along the two lengthwise grooves between the ridges, you would discover stacks of molecular disks called bases between the twisted vines, looking like thick, stiff leaves. These leaves would appear much alike, but on close examination you would see four different kinds: two bigger bases named adenine (abbreviated A) and guanine (G) and two smaller ones named thymine (T) and cytosine (C). Each base is made of nitrogen, hydrogen, oxygen, and carbon atoms and is flat, but because their atoms are arranged in slightly different ways, the bases have distinctly different outer edges. The bases are attached to the two twisted ridges by very short, stubby stems made of carbon, hydrogen, and oxygen atoms.*

A careful look at the bases themselves would reveal a considerable regularity in the pairs they form across the vines. Wherever an A comes from one stem, it is met by a T coming from the other strand; and if a G is coming from one strand, it always meets a C attached to the other strand. Anywhere you look along the double vine, you may find any sequence of leaves along one strand, but only two kinds of leaf pairs — G:C and A:T — would link the two strands. All in all, DNA would present itself to you as a pair of twisted vines, with G:C, C:G, T:A, and A:T pairs of inward-facing leaves coming off each vine to touch one another, making one complete turn of the double helix for every ten pairs of leaves. If you looked at the leaves on edge you could see through them, like venetian blinds.

If this is a text, where is the information? Information cannot be found in sameness. All the parts that are exactly the same from region to region on the DNA cannot hold information any more than a book of identical blank pages or a book filled with nothing but the same letter over and over. Where, then, is this double vine different?

Grab one of the two vines and climb on it; as you do, feel again the contours of the grooves that run between the vines. The flexible vines themselves are a simple repeat of oxygen, phosphorus, carbon, and hydrogen, the same everywhere along their length. But the grooves do not feel the same; they are wavy in a complicated way, like the grooves of a phonograph record. Looking inside the grooves, you see that the complex ripple is the surface of the outer edges of the four kinds of inward-pointing leaves, the four base pairs. Moving along the DNA, you feel the edges of the base pairs as a contour, and each is different, the way the milled edge of a quarter is differ-

---

* Names carry an evolved history of meanings. Adenine is a chemical that was first isolated from various glands: the Latin prefix for "gland" is *adeno*. Purified thymine too first came from a gland, the thymus. Cytosine was thought at first to come specifically from the cytoplasm, not the nucleus of a cell; guanine comes from guano (bird dung), which is rich in it.

ent from the smooth edge of a nickel. The sequence of ripples or contours you feel as you run your hand along the edges of a stretch of base pairs is the three-dimensional medium in which DNA stores its information. With a little bit of practice, you can recognize each of the four base pairs by feel, then transliterate this three-dimensional text by writing down the sequence in which the bases appear on one of the two vines as a string of English letters: AGCTA, and so forth.

The double helix of vines is a completely repetitive, stable structure, but the sequence of base pairs linking the vines needs not follow any pattern or order at all. Just the opposite: the sequence of base pairs in a DNA molecule is free to change without any effect on the structure of the outer double helix. Therefore, though at first the stack of leaves may appear to be an endless repetition of the four bases, you will discover as you write down the order of the base pairs that DNA is different from region to region. Sometimes you will feel a quite regular pattern of bumps in the groove as the base pairs run through a simple sequence over and over. Such a repeat is similar to a pitch pipe sounding a single note so that the instruments of an orchestra can tune up — it is necessary for the musicians, perhaps, but not the music. Sometimes, though, you will come upon a run of thousands of bases in which each of the four bases appears in approximately equal frequency but in no predictable order. Such a sequence of bases is precisely like the sequence of notes in a concerto or the letters in a book: no matter whether we look at a stretch of ten, one hundred, or one thousand notes or letters, the remainder of the sequence is not predictable from any of its parts. The absence of a repeating and therefore predictable pattern to the sequence of its bases is the sign that a particular stretch of DNA carries unique information. These are the genes.

If one strand can carry information in the sequence of its bases, why does DNA have two vines and a groove in which to feel the contour of a run of base pairs instead of just one vine with a sequence of bases appended, to be felt like beads on a rosary? The answer lies in DNA's double role as a text and as an inheritance: it has to have its own way of making copies of itself, to be delivered by parents to the first cell that will be their child, then copied thereafter by each cell so that it is present in all the cells of the body at all times in the life of the offspring. DNA's double-strandedness is its machinery for copying itself. In his firsthand account of the discovery of DNA's structure, *The Double Helix*, James Watson explained DNA's two strands with a wink: "Important biological objects come in pairs." By this glancing reference to the ubiquity of sex, he also meant to convey that each of DNA's two strands can be the source of an identical, new DNA molecule, so that the genetic information in one DNA molecule can be passed from one generation of cells — and individuals — to the next.

The clue to how this copying machine works lies in the special way DNA has of resolving what would seem to be an unnecessary ambiguity in the information it carries. A sequence of bases in a stretch of DNA can be read forward or backward; how does a cell know in which direction the DNA is correct? If you climbed along one of the DNA vines and came across a sequence like GGGAA, you could go backward and read it as AAGGG or you could jump to the other vine and feel the base-paired sequences CCCTT or TTCCC, depending which way you went. The base-pairing rules may predict a lot, but alone they leave the information in DNA ill defined. The ambiguity is resolved — and the copying mechanism revealed — by a closer examination of the vines. A real vine grows from the ground up and so has a natural top and bottom, but a molecule of DNA has no preferred direction. This is so for an unexpected reason: each strand of DNA does have an intrinsic direction, but the two strands always run counter to each other. Consequently, like the king of hearts in a deck of cards, a batonlike molecule of DNA can be twirled halfway around and still look the same.

Objects like DNA that retain their shape after a half rotation are said to have dyad symmetry. Imagine two trains on adjacent tracks leaving a station but traveling in opposite directions. At the windows of each train are passengers waving good-bye to the other train. If the trains were to stop and the passengers were to clasp hands across the platform, then the trains and their passengers, like the helical vines of DNA and their base pairs, would be in a state of dyad symmetry. A photograph taken from the air would show the trains linked by pairs of arms, and if there were no other landmark, one would not be able to tell which train was heading in which direction.

Look more carefully at those passengers reaching toward the hands of the passengers heading in the opposite direction. Their hands — palms facing in opposite directions — barely meet. As the tips of their fingers touch, each pair of hands forms an extended flat surface with fingers in contact running down the middle. To allow these flat surfaces to form, every person's elbow and shoulder must jut forward at a sharp angle from a window of one of the trains. Indeed, one could tell which end of the train held the locomotive, and which direction the train was about to go, simply by observing the angle made by each person's arm as it stretched to meet another's palm. Similarly, DNA's four bases all jut from their pair of vines at the same angle, meeting in between as a succession of flat base pairs. The base pairs have to be flat and stack perfectly, for if they puckered or met at an angle, DNA as a whole would not have dyad symmetry; rather, the direction of the bend or the angle of the base pairs would give DNA a built-in directionality, just as we could tell which train was which in our aerial photograph if the passengers on both trains were all pointing in the same direction instead of shaking hands across the platform. But the bases come off both vines at the same angle, so molecules that read the sequence of base pairs of DNA can first point themselves in the proper direction on either strand by recognizing this angle.

The base-pairing rules assure that the sequence on one strand will predict the sequence on the other. But because the molecules that read DNA in our cells always proceed from front to back, as seen from either DNA strand, and because dyad symmetry assures that front to back on one strand is the opposite direction from front to back on the other, special DNA sequences must first align these molecules so that they can only read the DNA of a gene in one of two possible directions, the so-called sense direction. By convention, the sequence of base pairs read in the sense direction is written from left to right. For instance, consider a DNA that has the sequence of base pairs G:C, C:G, A:T, T:A. The bases along one strand could be transliterated left to right as GCAT or right to left as TACG; the other strand could be transliterated as CGTA or ATGC. Let us say that the proper direction on one strand gives the sequence GCAT and indicate this by an arrowhead: GCAT>. Then the base-pairing rules tell us that the other strand would be properly transliterated as <CGTA. In order to follow a convention familiar to readers of the Romance, Slavic, and Germanic languages, transliterated DNA sequences are written from left to right, so the base-paired complement of GCAT> would be written as ATGC> rather than as <CGTA. This convention is awkward, but the arrowheads assure that the complementarity of two base-paired DNA strands is not forgotten.

Just as one set of molecules can use the fixed and constant elements of DNA's structure to align themselves for transliteration of the variable, informationally rich sequence of base pairs, other molecules see any double-stranded DNA molecule as the parent of two identical daughter molecules and carry out this doubling, called replication. The molecules of replication unzip the two strands of a DNA so that each can become the template for the construction of a new double-stranded DNA, whose sequence of base pairs will be identical to that of the original. Before a cell is fully prepared to duplicate its genes this way, it must stock up on a fresh supply of the four subunits of a new DNA strand. These subunits, called nucleotides, are made of a base already linked through a sugar to a phosphate group. Then, at the moment that one DNA is about to become two, the two strands separate.

Imagine a typesetter who places letters in a rack that will not take just any letters but will only accept a single, specific sequence of letters. The base-pairing rules leave no choice to the sequence of nucleotides that assemble as second strands on each of the two parental strands of a replicating DNA molecule. Each strand is the rack on which the nucleotides with proper bases line up to form a run of base pairs; each base on the original strand serves as a template for the addition of the proper nucleotide to the growing second strand. As new bases are appended to each strand to form a brand-new second strand, the base-pairing rules assure that each new double-stranded DNA molecule will be an exact replica of its parent.

All DNA, whether or not it encodes a gene, replicates this way. For most stretches of DNA, biologically useful information is held on only one strand, but the ubiquitous second strand, the way the two strands run in opposite directions, the law that G must pair with C and A with T — these three constraints mean that the information along an old strand of DNA is kept with fidelity by the new strand so that the informational strand — whichever it may be — is always assured of faithful replication as one molecule of DNA becomes two. The second strand is the minimum imaginable amount of extra molecular baggage necessary to make either strand's information self-replicating.

In 1944 the Austrian physicist Erwin Schrödinger predicted that the material of the gene would both be very stable and contain much information, despite the contradictory quality of these two attributes. Schrödinger captured the genetic material's exquisite requirement for both stability and meaning in an oxymoron, describing genes as being made of an "aperiodic crystal." By aperiodic, Schrödinger meant that the gene's material had to carry information. In this sense, any text in any alphabet must be aperiodic if it conveys information. A typical sentence, this one for instance, has information in proportion to its total length, because for every letter in the sentence, any of the other twenty-five letters might have been used to make any one of a very large number of alternative sentences, some of them with quite different meanings, others with no meaning at all. By calling the gene a crystal, Schrödinger meant that a gene needed the regularity and stability of crystals, the only other large things in nature that have lasted as long as life. The paradox was this: the gene had to have crystalline stability but it could not take the form of a crystal's purely periodic, informationally barren repeat. DNA — the base-paired double helix — completely removes the sting of paradox from Schrödinger's prediction. DNA is, precisely, an aperiodic crystal.

DNA's elegant architecture was worked out in the early 1950s. The first sign that it provided the solution to the problem of gene replication — albeit one that made the daring assumption that a gene's information resided entirely in its DNA — appeared as a brief poetical note to the April 25, 1953, issue of the scientific journal *Nature* by James Watson and Francis Crick. With this singular understatement, Watson and Crick opened the age of modern biology:

> The novel feature of the structure [of DNA] is the manner in which the two chains are held together by the . . . bases. . . . The sequence of bases on a single chain does not appear to be restricted in any way. However, if only specific pairs of bases can be formed, it follows that if the sequence of bases on one chain is given, then the sequence on the other chain is automatically determined. . . . It has not escaped our notice that the specific pairing we have postulated immediately suggests a copying mechanism for the genetic material.

∽

Some kinds of bacteria — small, single-celled creatures with a single molecule of DNA for a chromosome — have survived for billions of years. No one bacterium lives very long, but through DNA replication and cell division, the population persists. One bacterial cell becomes two by duplicating its chromosome, moving each new copy to opposite ends of the cell and splitting down the middle; each daughter bacterium then carries away one of the two new chromosomes. At every generation, the molecules of DNA replication use an old strand of DNA as a template for a new one. Consider what this means for these bacteria: among the ones alive today may be a rare bacterium that carries one of the two DNA strands from one of its ancient ancestors. In the extreme case, somewhere on Earth might be a bacterium with DNA made up of two strands of absurdly different ages: one made twenty minutes ago and the other almost as old as the planet itself.

Even if such a rarity exists, we have no way of discovering it; the base-pair sequence in the ancient DNA strand would not be chemically distinguishable from that of any other bacterium of the same species. The planet's surface has changed many times over, but DNA and the cellular machinery for its replication have remained constant. Schrödinger's "aperiodic crystal" understated DNA's stability: no stone, no mountain, no ocean, not even the sky above us, have been stable and constant for this long; nothing inanimate, no matter how complicated, has survived unchanged for a fraction of the time that DNA and its machinery of replication have coexisted.

But nothing alive is ever perfect. Though the backbones of DNA strands may last for a very long time, the information within DNA can change by accident at any time. The rigid base-pairing rules, which enable information to be copied from one DNA double helix to two, can also fix in place any error that occurs. Sometimes such an error can be calamitous. The word "mutation" was coined in 1901 by the Dutch botanist Hugo De Vries to describe the sudden appearance of a new variant form or behavior that did not go away but rather bred true in successive generations. The best-known single mutation is the one first seen by one of the founders of modern genetics, Thomas Hunt Morgan. In 1910 Morgan reported that he had found a single white-eyed fruit fly among the tens of thousands of little red-eyed fruit flies that his laboratory maintained. In short order he showed that this fly was genetically different from normal; that is, its descendants would sometimes — but not always — have white eyes as well, and, like color blindness in humans, this mutation affected males much more frequently than females. Descendants of the first white-eyed fruit fly are still giving their cells, chromosomes, and DNA to scientists in hundreds of laboratories throughout the world today.

Mutations are changes in the sequence of bases in the DNA of a gene. Even a single base pair, if copied or repaired wrong just once, can completely change the meaning of a stretch of DNA and preserve that changed meaning in all subsequent generations of that DNA. These changes are often harmful, and our cells have many ways of preventing them. For example, consider the possible consequences of a summer spent enjoying long, lazy days of sunshine at the beach. Your skin responds by darkening to protect the DNA in the dividing cells beneath the skin from the errors an invisible fraction of sunlight called ultraviolet can introduce. Everyone's skin is made of many kinds of cells. The cells whose clumps we shed as dandruff are cross-linked to form the waterproof layer that keeps us from melting in the rain. Beneath these dead cells lies a single layer of living, dividing cells, whose daughter cells push up to replace the dead and dying skin cells as they are shed. The layer of live cells also contains a number of pigmented cells called melanocytes, from the Greek for "black cells."

As each melanocyte fills with tiny sacs of black pigment, it hands these off to brand-new skin cells. The skin cells of populations raised far from equatorial sunlight receive relatively few sacs of black pigment from their melanocytes, so their skin looks beige or pinkish from the blood vessels beneath; the skin cells of populations close to the equator receive many packets of pigment, so their skins range from brown to black. In all people, melanocytes respond to sunlight by dividing and by increasing the number of pigment granules they place in skin cells, further darkening the skin to shield the DNA of the dividing skin cells.

When the melanocyte shield fails, the ultraviolet portion of sunlight can lead to changes in the DNA of a cell that has yet to divide, introducing an error in one or more of its genes. In a sequence that contains two A:T base pairs in a row, the two thymines lie so closely on top of each other that the energy of ultraviolet sunlight can form a bond directly between them, called a thymine dimer. The molecules that carry out DNA replication cannot get past this unexpected compound, so usually replication ends at that point, and a skin cell dies. Sometimes a cell can suffer a self-inflicted wound: it can inadvertently change its DNA sequence as it tries to repair the damage done by ultraviolet light.

Repair enzymes are always on the job, snipping into DNA's backbone and removing mismatched or damaged bases, replacing them with new, proper bases, and then sealing up the loose ends on the damaged strand. If the repair of a thymine dimer is successful, then the skin cell and its descendants will be normal. But if the repair enzymes responsible for fixing a thymine dimer snip it out and reseal the strand without first inserting two good thymines, the "fixed" strand suffers the loss of two Ts, a change as consequential to the DNA's meaning as to the meaning of a sentence in which "tattle" becomes "tale." If the deletion is faithfully copied into a full double-stranded DNA during DNA replication, it will then be propagated in one of the two daughter DNAs and its descendant cells. The change in base sequence would no longer be seen by the repair systems of daughter cells as an error to be fixed, so it would be propagated thereafter as long as the cells in which it resided were viable. In this way, a change in base sequences, or a mutation, can be locked in DNA for many generations.

One rare but dangerous consequence of such a sunlight-induced error is a cluster of rapidly growing variant cells, which we call skin cancer; when the cancer grows from a mutated melanocyte, it is called a melanoma. Often, the cells of tumors — including tumors of the skin — show a rearrangement of chromosomes. Not surprisingly, then, agents we know as causes of cancer, such as the tars in cigarette smoke, are capable of damaging and rearranging our chromosomes. Sunlight, chemicals, and other forms of radiation can also cause complete breaks in DNA. The faulty repair of such breaks can also massively rearrange the sequence of DNA in a genome. Large enough rearrangements can actually change the gross appearance of chromosomes, as pieces of DNA millions of base pairs long break away from one chromosome and attach to another. When such rearrangements, called chromosome translocations, occur in the cells that will merge to begin a developing embryo, they are often fatal; where they are not, they will often result in serious illness from birth on.

Mutations are rare if we measure their rate of appearance relative to the lifetime of an organism. But life is old, and in the course of its existence some new sequences of base pairs have survived for a long time; others have been lost quickly and forever. While some, perhaps most, mismatches of base pairs during DNA replication generate neutral or lethal mutations, beneficial mutation is the underlying cause of life's diversity. If DNA replication were perfect, without error, life would have died out long ago from its failure to adapt to the fluctuations in temperature, atmosphere, and water level the earth has seen over the ages. With slight but continuous mutation, however, the descendants of some organisms have been able to survive myriad environmental perturbations to become the millions of species of flora and fauna we recognize today.

DNA's wasteful but so far successful strategy for surviving environmental stress and competition through less than perfect replication is the fuel that powers Darwinian natural selection. For example, where stereotypical altruistic behavior has been examined carefully, it has been found to enhance the survival of close relatives of the altruistic individual. Kin selection, as it is called, takes place at the cost of one individual's life but raises the probability that some of that individual's unique DNA sequences will survive, passed on by relatives. Since the sequences of DNA in the chromosomes of any organism are the sole vehicle for transmitting viability through time, we must grapple with a depressing, reductionist summary of natural selection: all life is DNA's way of making more DNA. To those of us who choose to make it so, life is more than that; but we cannot call on biological justifications for our choice.

Without exception, members of all living species carry shared DNA sequences, a fact that is consistent with the bracing notion that a single DNA-based life form was the ancestor of all living things — the Big Birth theory, as it were. Just as paleontologists sometimes can accurately date the fossilized remains of an ancient species using geological markers, molecular paleontologists can determine the rates at which DNA sequences for the same gene have diverged over time. They have found that many groups of ancient genes have diverged very slowly, changing about one to ten base pairs per million years. From such rates, called molecular clocks, they can propose how long it has been — usually in millions to hundreds of millions of years — since the last common ancestor between two living species was itself alive. Given the difficulties of getting exact dates and clear identifications for ancient remains and the risks of idiosyncrasy in the choice of genes to analyze, it is remarkable how often molecular and field paleontologists can roughly agree on the age of a long-vanished species.

Darwin assumed that natural selection would generate a slow but smooth appearance of new species, but in some cases the fossil record suggests that species may be stable for very long periods of time, only to be supplanted in short bursts of species proliferation. These often follow hard upon cataclysmic extinctions of many species. Niles Eldredge and Stephen Jay Gould have a name for this staircase variant of Darwin's smooth ramp of speciation: punctuated equilibrium. Certainly the lifetime of some living species can be very long: the fairy shrimp has been around for more than two hundred million years, or about one twentieth of the entire time life has been present on Earth. The oldest fairy shrimp fossils look slightly smaller than living samples but are otherwise indistinguishable from them. Perhaps the stability of their environment over unusually long periods has contributed to the longevity of the species by forcing the loss of what must have been, over all those years, a number of individuals bearing a very large collection of different spontaneous mutations.

Rich in variations on the theme of life as DNA has become, most DNA sequences never existed and never will. In order to appreciate the abundance of untapped possibilities in the DNA of our chromosomes, consider a very short sequence of bases that runs 17 base pairs long. A small calculation reveals that there are 4^{17}, or approximately 17 billion, different sequences of seventeen base pairs. Since there are only about three billion base pairs in any one set of twenty-three human chromosomes, any particular stretch of seventeen base pairs should appear at random no more than once in any human genome. There are about six billion people alive today; each one can be identified by a single unique string of seventeen base pairs, with more than that number of unused sequences left over.

With so many sequences possible, the discovery that our chromosomes carry many long stretches of DNA with almost identical sequences comes as a bit of a surprise. Names help to explain multiple mentions, or families, of DNA sequences. The six billion people on the planet can get by with rather short names. In no more than seventeen letters we can usually get both a first and second name and a middle initial. Since the purpose of names is to distinguish us from one another while retaining information about our family origins, the uniqueness of names is consistent with the improbability of repeating two long stretches by accident. Only rarely do two unrelated people have identical full names, and when the situation arises, we can be pretty sure it is not entirely a random occurrence but the result of one or both of two possibilities: either the two people are related or the name is quite common, or both. Similarly, when human gene sequences are deciphered, some that are a few dozen to a few hundred bases long are present throughout the human genome, in thousands or even hundreds of thousands of copies. Other sequences are found in sets of genes that have a common ancestry or a common function or both. In these cases, we assume that the multiple sequences arose by duplication and reduplication from a single sequence, not by the vanishingly small possibility of coincidence.

Looking at the richness of life on Earth today, it is hard at first to believe that natural selection — which permits the survival of some but not all randomly occurring sequence differences in DNA — is responsible for so many elegant designs. But every mutation may be a new design, and each mutation that survives must make sense in its own context before it can serve as the new baseline for the next mutation. In this way, a series of changes will accumulate over time in remarkable mimicry of intelligent design. We can get the same effect by selecting from a set of slightly variant words, all of which make sense. Consider the following statement, which ought to be clear to every reader at this point: "A base in DNA encodes the data of a gene, storing information in the text of life." We can also get from the word "base" to the word "life," passing through the words "data," "gene," and "text," in a series of single-letter mutations, all of which generate new words that make sense. One of several ways is: *base*> bale> dale> date> *data*> date> Dane> cane> cant> cent> gent> *gene*> gent> tent> *text*> tent> cent> cant> cane> lane> line> *life*. My sentence using the five words is reasonable and logical. In comparison, the sequence of words in the selective sequence makes no sense, although each word has meaning. Both reach the same words in the same order. In the same way — locally smooth while globally random — the natural selection of a meaningful minority of changes in DNA generates spectacularly complex structures, which seem in retrospect — but only in retrospect — to be the result of an intelligent plan.

~

Earth spins in a void as cold as any the Universe allows, which presents a time paradox. No matter how hot Earth once was, it should have cooled, skin first, rather rapidly once it formed, like a potato removed from the stove. Darwin was unable to explain how the Earth's surface could have remained temperate for long enough to allow life to evolve. The answer to this puzzle was provided about a century ago from a wholly unexpected source, one full of resonance in our atomic age: the discovery of radioactive elements and of their capacity to release vast amounts of energy from the disintegration of their atomic nuclei. The core of our planet is full of radioactive materials, and the heat they slowly release has warmed its surface over the billions of years since it formed. More recently natural selection, building on the mutation of DNA, broadened the spectrum of living DNA sequences until they arrived at those encoding us.

We — in the briefest instant as measured on this scale of time — have now learned how to transmute both atoms and DNA for our own purposes. While natural selection continues, and will continue while life on Earth remains, it will never again be entirely constrained by random mutation nor entirely protected from self-induced catastrophe. If all other species can be born, live a certain amount of time, have offspring, and one day die, then we must consider the likelihood that this will be the fate of our own species. The big brains that have brought us to consciousness, memory, and the dream of immortality have given each of us a deep sense of individuality, a sense that makes it difficult for us to grasp the notion that our species — not any one of us, nor any one family, race, religion, or nation — is the smallest unit of survival through natural selection and that its survival is anything but assured.

2

~

CHROMOSOMES
AND CANONS

HUMAN GENOMES ARE PLENTIFUL but fleeting. At the conception of every child, a distinctly different human genome is bound and issued for the first and only time. There are as many human genomes as there are persons: six billion drafts of the human DNA text are clustered over every part of the planet not covered by water. Each human genome differs from all the others because — while each is made of the same set of about a hundred thousand genes — a gene need not be restricted to a single version of its precise text. Indeed, many genes denote themselves in different versions, called alleles (pronounced ah-*leels*; from the Greek for "of one another").

To see how different versions of a gene coexist, we can compare the text of the human genome to another ancient (or at least very old) text, the New Testament. First set down in Greek, the New Testament has been translated and retranslated into English for the past five centuries. Each translation, from the Tyndale Bible of 1525 to the 1960 Revised Standard Version, has held great meaning for millions of faithful Christians. But though they are all the New Testament, each version is different. For example, here are the six versions — alleles — of a single sentence from the Book of James, Chapter 4, verse 5:

TYNDALE, 1525 *(the first English Bible):* "Either do ye think that the scripture sayeth in vayne. The sprite that dwelleth in you, lusteth even contrary to envye."

GENEVA, 1562 *(Shakespeare's Bible):* "Do yee think that the scripture saith in vaine, the spirit that dwelleth in us, lusteth after envie?"

DOUAY, 1582 *(the first Catholic translation):* "Or do you thinke that the Scripture saith in vaine: To envie doth the spirit covet which dwelleth in you?"

KING JAMES, 1611: "Do ye think that the Scripture saith in vain, the spirit that dwelleth in us lusteth to envy?"

AMERICAN STANDARD VERSION, 1901: "Or think ye that the scripture speaketh in vain? Doth the spirit which he made to dwell in us long to envying?"

REVISED STANDARD VERSION, 1960: "Or do you suppose it is in vain that the scripture says, 'He yearns jealously over the spirit which he has made to dwell in us?'"

With the help of a good concordance we find that within each allelic version of James 4:5 is a reference to an earlier work: the phrase "the spirit he planted within us" comes from "the breath of life," the divine spark in each human being first mentioned in verse 2, line 7, of Genesis, one of the five books of the Torah at Jerusalem's center. Genes, too, carry earlier sequences from other, earlier genes; deriving the full historical meaning of a gene can begin only when a full concordance of the genome is available.

A comparison of these six biblical sentences shows us that this single line from James has no single, perfect translation. Since the Greek is ambiguous about whether "the spirit" is the subject or object of the verb "to yearn," the sentence was given two quite different meanings by its last two translators. Does this mean that the Bible read by Shakespeare was "incorrect"? Not at all. And what is true for sentences in any important book — but especially for a book that has been important for a long time and has gone through many editions, reprintings, and translations — is also true for genes: meanings will multiply with time, and no single allele is ever going to be the sole "correct" version.

The process of natural selection confirms this point. If two or more alleles work equally well in terms of fecundity and the survival of offspring, no single allele is "normal" or "bad." It is therefore not surprising that any two healthy people are likely to carry different, distinctive alleles for many genes. Blood type, for instance, is largely determined by which of three normal alleles — A, B, or O — a person carries. Until the discovery of blood groups A, B, and O at the turn of the century led to safe blood transfusion, these alleles were a set of distinctions without a difference.

The existence of several alleles for a single gene is the basis of human individuality: a person can inherit two alleles for any gene but can only pass along one of these to a descendant. The inheritance of one or the other allele is a matter of chance at the moment of conception; as a result, no child is the exact genetic copy of either parent. Multiple alleles have been found for almost half the human genes studied so far; just as a vast number of hands can be dealt from only fifty-two cards, the number of possible assortments of alleles is large enough to assure that no two people will ever be born by coincidence with exactly the same genomes.*

Any change in a DNA text that generates a new allele — even a change as simple as the addition, removal, or substitution of a single base pair — is a mutation. Some mutational misprints of DNA's less than perfect copying machine leave a gene's meaning intact, but others can make its message dangerous or even lethal to the organism. Single base-pair mutations in the genes that encode hemoglobin — the oxygen-binding pigment inside red blood cells that gives them their color and us the oxygen we need to convert our food to energy — can have quite dramatic effects. One such typo produces the sickle-cell mutant allele of a hemoglobin gene, which makes abnormal hemoglobin molecules that lock into stacks, distorting the normal bagel shape of a red blood cell into a pointy-

---

* If no more than three hundred of the hundred thousand human genes — instead of the tens of thousands we expect — were found to be multi-allelic, then there would be at least 2^{300}, or 10^{100}, different human genomes possible. Compared to this absurdly large number of possibilities, the fraction of possible genomes that have actually existed is vanishingly small: only 10^{80} atomic particles are said to make up the entire known universe.

ended croissant. The bent blood cells pile up in the tiniest blood vessels, shoving as ineffectually as a crowd of tourists trying to get through a revolving door. A person who inherits two sickle-cell alleles for the hemoglobin gene — and therefore no normal alleles — will have sickled red blood cells and will suffer from crippling bouts of anemia, tiredness, and internal organ damage. But if a person's genome includes one normal allele and one sickle-cell allele for the hemoglobin gene, enough normal hemoglobin will be made to allow for a nearly normal life.*

Whether normal or damaging, alleles are gifts we receive from our parents and give to our children. Both parents inherit twenty-three pairs of chromosomes, but because sperm and egg cells are haploid — meaning they carry only one copy of each gene — each parent can give only one half of each chromosome pair to a child. To a parent, this means that as much genetic information will be cast off as passed on each time a child is conceived. To a grandparent the message is even more severe: because a baby's cells can have only one of each parent's two alleles, the alleles from two of its four grandparents cannot be included in the genome of a child. Worse yet, the choice of which allele is sent on to a child — mother's mother's, mother's father's, father's mother's, or father's father's — is entirely random, and the choice usually differs from gene to gene and child to child.

Parental chromosomes are not delivered unchanged from the bodies of a mother and father into the sperm and egg that fuse to make a child. Instead, as sperm and egg cells are made — the process is called meiosis — pairs of chromosomes carrying the same genes lie next to each other and exchange DNA. Alleles of genes that lie next to each other are then switched

---

* Why has the sickle-cell allele not been lost? While a double dose of the sickle-cell allele usually results in early death, a single allele confers some protection against malaria. The tiny, single-celled animal that causes malaria lives for a time in red blood cells and finds sickle-cell hemoglobin inhospitable. People with one sickle-cell allele therefore suffer fewer and less severe bouts of malaria than those with two alleles for normal hemoglobin. Not surprisingly, carriers of one sickle-cell allele — as well as children born with sickle-cell anemia — are both common in the parts of Africa, the Middle East, and South Asia where malaria is endemic.

about. Once the new sperm or egg cell is created, a gene on one of its chromosomes may be represented by an allele that came from the parent's mother, whereas the next gene may be an allele that originally belonged to the parent's father.

All alleles come from a grandparent; this is the genetic basis for the closeness of a family's resemblance. Traits that involve the expression of great numbers of genes are the clearest indicators of the role of inheritance in the choice of alleles. Facial appearance is the most obvious one: overall, children do resemble their parents and grandparents more than their other relatives. Nevertheless, every baby is an assemblage of choices from the chromosomes of its parents. The process of shuffling the alleles from each parent's two parents — called recombination — assures that every chromosome that goes into egg or sperm will carry an assortment of alleles that never existed in the chromosomes of any of the new child's four grandparents.

At the level of individual genes, a baby is no more related to its parents than it is to anyone else who carries the same pair of alleles for a particular gene. This counterintuitive notion kills all hope that studying the sequences and mechanisms of action of a small number of human genes will allow us to understand the wellsprings of human individuality. While it is becoming possible sooner and sooner after conception to determine which alleles of a gene a fetus has inherited, recombination during meiosis randomizes the allotment of grandparental alleles to each child in a way that cannot be controlled by any technology we can envision. Recombinant DNA — produced at every meiosis — is not so much a human creation as an ancient and necessary step in the development of a unique genome for every person.

For any gene, the particular pair of alleles in a person's genome is called the genotype. When two people differ in form because they carry different genotypes, their genotypes are said to result in different phenotypes. (The word "phenotype" contains the notion of appearances as distinct from their underlying realities; it comes from the same Greek root as "phantom.") Phenotypes like eye color show themselves in the outward appearance of a person; others, like blood type, are hidden beneath the skin. The distinction between inherited alleles and their visible consequences — between genotype and phenotype — raises a deep question about DNA as a text: why do we need the word, and the notion, of a phenotype at all? In principle, if all forms are encoded in DNA, should not every outward phenotype imply a single, distinct genotype? Why not, then, simply speak of genotypes and drop the redundancy of the phenotype, the phantom of appearances?

The answer is simple but surprising: a crucial ambiguity allows different genotypes to generate the same phenotype. This ambiguity arises whenever either one or two copies of an active allele produce the same phenotype: a person with the phenotype of type A blood may be carrying either one A and one O allele or two A alleles. Usually, an allele that has no effect on the phenotype unless it is present in both copies is inactive. For example, while the A and B alleles are responsible for two slightly different chemicals on the surface of a red blood cell, the O allele is silent and inactive, neither adding nor subtracting anything. In its silence it offers no barrier to the function of a second allele of either the A or the B type, so genotypes of AO and BO generate phenotypes indistinguishable from AA and BB, respectively. An active allele producing the same phenotype, whether present on one chromosome or both — like the one for blood type A — is called dominant. The phenotype resulting from the presence of two inactive alleles — type O blood — is called recessive, and the inactive allele itself — like the one for blood type O — is often called a recessive gene.*

Though we cannot tell by looking at a person with a dominant phenotype whether the gene for that phenotype is present on both chromosomes or only one, the distinction may have consequences in future generations. Take eye color. The same pigment cells that give the skin a tint of beige or brown or black also sit at the back of the iris. When the appropriate allele commands them to hand off pigment sacs to the cells of

---

* If a person inherits a matched pair of alleles for a particular gene, he or she is said to be homozygous for that gene (from the Greek for "identical" and "joined together"); a person inheriting a dominant allele from one parent and a recessive allele from the other is said to be heterozygous for the gene (from the Greek for "different" and "joined together").

the iris, we see a brown- or black-eyed person. When an inactive allele cannot give this command, we see a blue- or green-eyed person, because by itself the unpigmented iris, like a swimming pool, reflects the bluish colors of light and absorbs the red. Although more than one gene contributes to the phenotype of eye color, the dominant allele of the major gene puts pigment into the iris, while the recessive allele of this gene does not. Blue-eyed people are therefore certain to have inherited two recessive alleles, while brown-eyed people may have inherited one or two active copies of this gene; they cannot tell which by looking each other in the eye.

Two blue-eyed people can be fairly certain that all their children will share their eye color, since all four possible alleles of the eye pigment gene in all of their sperm and eggs are going to be silent. But two brown-eyed people who hope all their children will inherit their brown eyes cannot predict that this will always happen. If both parents have genomes containing one dominant and one recessive allele of the eye pigment gene, they stand a good chance of having a blue-eyed baby. After meiosis puts the silent, recessive allele for eye color in half of the father's sperm and in half of the mother's eggs, a quarter of their children are likely to have blue eyes.

Something esoteric is going on here: the silence of the recessive allele need not be permanent. Recessive and dominant alleles are just slightly different sequences of DNA, and both are equally stable through any number of generations. A recessive allele may be passed from parent to child again and again, silent but persistent. Then, when two random parental throws of the allelic dice bring one parent's recessive blue-eye allele together with the other's recessive allele, blue eyes surface, as blue as any other, even though the child's genealogy may show nothing but brown-eyed ancestors. If such parents are surprised by the eye color of their baby, it is because they did not know that each of them was carrying a silent allele in addition to the dominant one that gave them their brown eyes.

The persistence of recessive alleles also belies the common notion that our form — the shapes and colors and traits that make us who we are — is inherited by some sort of blending of ancestral fluids. The womb is not a genetic mold into which a child's future constitution is poured; there is no mixing of two inheritable liquids, no "blood," no "blue bloods," no "bloodlines." Children are assembled as a collection of discrete, randomly assorted, stable, dominant and recessive ancestral alleles. The difference in these two notions of form is important. One can imagine an undesirable trait encoded in the genetic fluid of an ancestor eventually being diluted out simply by the passage of generations. But the stability of recessive alleles means that the silent texts in a person will not necessarily be diluted away in the children and grandchildren arising from the marriage of that person to someone of the "right blood." Well bred as some of us may wish to think we are, the genomes of each of us and each of our children remain mosaics containing some large number of inescapable, undilutable recessive alleles. Single copies of recessive alleles, silent in the presence of their dominant alternatives, keep us from being able to judge all genotypic books by their phenotypic covers.

We have had evidence of the difference between genotype and phenotype since 1863, when Gregor Mendel described the hidden inheritance of recessive alleles and elevated genetics from an accumulation of farmers' habits to a quantitative science. Mendel carefully recorded the phenotypes of many generations of experimentally mated pea plants. In all, he followed seven pairs of alleles and found that in each case one allele — smooth peas, say — was dominant and the other — wrinkled peas — recessive. A cross between parents carrying one of each allele always produced a mixture of offspring, with about one in four showing the recessive phenotype. Since genes that are on the same chromosome travel together from generation to generation unless recombination separates them, it is quite remarkable that he obtained such clear statistical data for each pair. Apparently, although he could not have known it — chromosomes had not yet been described when he did his work — the species of pea he used had seven chromosome pairs, and each gene he studied happened to be on a different chromosome pair.

The wrinkled pea allele itself has an interesting story. The strain arose when a foreign piece of DNA inserted itself into the middle of a gene that makes starch, rendering it inoperative. Starch is one of the foods a pea seed stores for later use by the growing seedling, and storing starch is a major job for the pea: many genes encode the molecules that link sugar molecules into the great chains and webs we see as starch granules. If both alleles of one of these genes are disabled, a pea's capacity to make starch is ruined. Starch will hold water and swell — think of tapioca pudding. Simple sugars, though sweet, will not hold water as starch does. Failing to convert sugar into starch, doubly recessive peas show not one but two phenotypic differences from dominant, smooth peas: they do not stay plump as they grow in the pod, and they taste much sweeter than smooth, starchy peas. The sweetness of wrinkled peas no doubt made them desirable to cultivate, which is probably why Mendel had true-breeding recessive strains of wrinkled peas for his work.

Mendel was a farmer, priest, and schoolteacher in Brno, a city then in the Moravian portion of the Austro-Hungarian Empire and now in the Czech Republic. He must have been a person of great observational skill, almost unimaginable patience, and great good fortune. A student at the University of Vienna for two years, he knew the work of other European botanists. How pleasantly unexpected that this celibate scientist, and not any of his more worldly peers, had the wit and perseverance to discover the mechanism that governs inheritance from pea plant to pea plant — and from parent to child. And how fitting that a priest should have found that this mechanism is built on a text that can remain silent for any length of time but then speak out again with full force.

∾

The most dramatic phenotypic difference determined by the presence of a single allele is sex: boys are male like their fathers and grandfathers; girls are female like their mothers and grandmothers. This pattern of inheritance suggests that the choice between male and female in a developing embryo, like the choice of wrinkled or smooth in a developing pea seed, is underwritten by a choice from a single pair of alleles. This is true, but the gene that determines the sex of a child differs from other pairs of alleles in one important way: it has its own chromosome. While women have twenty-three pairs of matched chromosomes, including one pair called X and X, men have twenty-two pairs plus one mismatched set, called X and Y.

The normal allele for male sex on the Y chromosome is dominant: in its presence an early embryo grows into a male; embryos that lack it grow into females. The biblical story of Eve's creation from a part of Adam's body is thus backward; all human embryos begin by looking female. The action of the sex-determining gene on a male embryo's Y chromosome transforms an initially female body into one that will belong to a little boy. This gene — called SRY — acts by turning on a cascade of other genes, beginning with the ones that form the testes, in a precisely timed fashion throughout the developing male embryo. The human genome must contain two interleaved texts, one to be read only in male cells and the other only in female cells; both are present in interdigitated form in the DNA of all cells of both men and women. How much of what we assume is a common genome is in fact held in reserve for one sex? It speaks to the subtlety of our genomes that all of us have DNA sequences our bodies will never read — sequences that can be read only by cells in a person of the other sex.

The cells of girls as well as boys use the information on only one X chromosome. Early in the development of a female embryo, one X chromosome in every cell is rendered permanently silent by a chemical coating called methylation. Since the inactive X chromosome is chosen at random in each of its cells, the female embryo grows up to be a mosaic of cell patches expressing different X-associated alleles: a tabby coat, for example, is seen only on female cats. Short regions of many other chromosomes are methylated even earlier, during the formation of sperm and egg. This localized methylation — called imprinting — can make two identical alleles behave differently in the developing human embryo, and the embryo's normal development requires both the maternally and the paternally imprinted versions of each gene.

The inheritance of a Y chromosome is almost always

sufficient for the inheritance of a male body, and exceptions confirm the dominance of the Y gene. Rarely, a human egg is successfully fertilized by a mismade sperm bearing an X chromosome but also — stuck onto another chromosome — the critical piece of DNA from a Y chromosome. A person who develops from such a fertilized egg will be physically male even with two X chromosomes in each of his cells. Conversely, a person born from an egg fertilized by a sperm carrying neither the X nor the Y chromosome will have only a single maternal X chromosome in each cell but will be physically female. Normally, though, the choice between the conception of a girl or a boy is simply a matter of whether an egg is fertilized by a sperm carrying an X or a Y chromosome.

About half of the sperm in a normal man carry his X chromosome and half his Y. If X sperm and Y sperm each fertilized eggs with the same efficiency, half of the children born should be of each sex. In fact, about 5 to 10 percent more boys than girls are born, suggesting that the Y sperm may have a slight competitive advantage or that a male embryo does slightly better in the womb. However, once born, males are at greater risk of dying. By the time people are old enough to have children, the ratio of mothers to fathers has settled back to about even numbers, then keeps dropping with increasing age. As anyone who has visited southern Florida knows, there are many more retired grandmothers than grandfathers. Perhaps this is partly the consequence of culture or of the hormonal balances of men and women, but recessive alleles on any man's X chromosome certainly take their toll as well. While a boy and his sister may both inherit the same recessive allele with their mother's X chromosome, the girl can — but the boy cannot — look to a second X chromosome from their father to provide a functional allele of the gene. Boys will therefore have a higher probability than girls of displaying the consequences of such X-linked alleles as red-green color blindness and hemophilia.

Soon after Mendel's discovery of recessive alleles was absorbed by the scientific community — it took more than thirty years — alleles that were neither recessive nor fully dominant began to turn up. The inheritance of one recessive allele and one of these incompletely dominant alleles produces a third phenotype, different from those generated by two normal or two mutant alleles. For example, a cross of red and white snapdragons produces pink flowers. The white-flowered snapdragons have two copies of a silent allele for a gene that normally produces red pigment, and red-flowered plants have two copies of an active, pigment-producing allele. But as it happens, one active allele for the red pigment is not sufficient to fully saturate the petals with color, so we can see at a glance that pink flowers have one allele that produces the pigment and red flowers have two. The "pink snapdragon" phenotype, with one allele functional but unable to serve for two and the other silent or missing, can sometimes be less than completely healthy. For example, even in the presence of normal hemoglobin encoded by a single functional allele, the sickle-cell allele produces enough abnormal hemoglobin to cause difficulties at high altitudes, where oxygen is in shorter supply.

Even when a functional allele is fully dominant, the inheritance of only one copy instead of two can lead to problems, especially when that remaining copy must function throughout the body in order to maintain an essential aspect of normal form. For example, the family of genes that works together to keep cells in our body from dividing out of turn also keeps tumors from arising in our bodies. One member of this gene family, p53, has been studied with particular care. When mutations — like the ones caused by the ultraviolet component of sunlight — damage both alleles of p53 in the chromosomes of any cell of the body, that cell and its descendants can begin to divide in an uncontrolled way until they grow into a malignant tumor. Biologists know that p53 plays a major role in keeping tumors from sprouting, since both copies of the gene are damaged or lost in a majority of human tumors.

Normally, when a person inherits two functional alleles of the p53 gene, the second allele of p53 provides the body with a valuable redundancy. Mutations in a tissue cell may inactivate one of its p53 alleles, but the cell will not become a tumor because the remaining allele will still produce enough p53 to keep it in check. A person born with only one functional allele of p53 will appear healthy at first, because one p53 allele can hold all the body's cells together. But any time a cell anywhere in the body loses its sole active p53 allele through mutation, a tumor is likely to grow. The phenotype of people with only one functional p53 gene is therefore abnormal in an insidious way: at all times they are highly susceptible to developing a tumor. Some families with this propensity share the Li-Fraumeni syndrome: its victims develop multiple tumors early in childhood. The phenotype of increased susceptibility to cancer reappears in Li-Fraumeni children as frequently as if it were a dominant allele: a man or woman with one damaged p53 allele will inevitably put that allele into half his sperm or half her eggs, so about half of his or her children will inherit a tendency to develop cancer even if their other parent has two functional p53 alleles. P53 is not the only gene linked to the dominant inheritance of cancer susceptibility in this way. Another, RB, is even nastier: when active from conception in only a single allele, it gives children tumors in their eyes before their first birthday.

~

No one wants to inherit a disease, and no one wants to pass on the susceptibility to one to their children. It would be handy to be able to take the sperm or eggs from a person and select only "desirable" sperm to fertilize a "desirable" egg, so that every baby was healthy. But recombination and recessiveness together mean that we cannot prepare a cell to become the sperm or egg we wish. Alleles are tucked away in the chromosomes of sperm and egg cells; like an electron resisting any effort to pin down its position and direction of movement, any allele in an egg or sperm would escape from our grasp as we opened up and killed the cell that carried it. Along with the recombination of alleles and the random assortment of chromosomes during meiosis, this makes it hard to imagine how prospective parents will ever be able to choose an egg and sperm free of any particular pair of recessive alleles that might make for an unhealthy baby.

While the selection of alleles before conception seems unlikely, DNA analysis of blood or biopsy cells can reveal which alleles a person received at conception. The equipment for doing this requires a very small amount of tissue, in some cases no more than a single cell. With this capacity in the hands of a wide range of professional people, from molecular biologists to physicians to lawyers to the FBI and the army, two distinctly different terrains lie ahead. Both involve interpreting DNA from people at risk, and both are likely to be full of legal, ethical, and political potholes. Since it has never been a habit of scientists to sit with an idling engine at a turn in the road, for better or worse both are likely to be well traveled.

Down one road, the tools of DNA analysis are being used to detect and recover defective alleles from the tissues of persons already afflicted with genetic diseases. The goal is to learn how a specific allele differs from normal, then to create drugs that might block or reverse the effects of the abnormal alleles in people — or embryos — who would otherwise be vexed with one or another inherited disease. This calls for some tough choices: every time the DNA sequence of a defective allele for an incurable disease is found, potential carriers may be obliged to learn of their gloomy inheritance before research has given them any way to ameliorate it.

Scans of disease-related alleles, especially when they run ahead of treatments for the disease, often have serious and unintended consequences, damaging the lives of those they were designed to help. In a classic example from the early 1970s, the U.S. government mandated that certain arbitrarily chosen and poorly defined groups of healthy citizens — prisoners, African Americans, Hispanic Americans — would have to be tested for the presence of a sickle-cell allele. Although the stated aim was to help reduce the number of children born with sickle-cell anemia, no counseling or any medical care was tendered. After a few years of considerable outcry, the government ended the mandatory program.

The road to early detection started out a few decades ago as a set of assays of phenotype that could be carried out on children just after birth. For instance, a drop of blood must be taken at birth from every child born in some states — California comes to mind — and tested for the presence of chemicals that signal the inability to properly digest foods containing the common ingredient phenylalanine. This inherited disease, called phenylketonuria or PKU, profoundly retards children who eat foods containing phenylalanine in their first ten years, while the brain is still growing. Magically enough, PKU children can have a normal life if they are spotted immediately and given foods with little or no phenylalanine until they are about ten years old. Spared by proper diet in childhood, many women carrying two PKU alleles have reached childbearing age. Providing they drop their dietary intake of phenylalanine around the period of pregnancy and make sure that their children are tested for PKU at birth, both mothers and children can be protected from the consequences of their mutant alleles.

Tay-Sachs disease is another syndrome that ends in retardation and death, but we have no cure for it. It is the lethal consequence of an unlucky union of two silent alleles for a gene that helps complete the outer surfaces of growing nerve cells. The nervous system of a person who inherits either one or two functional alleles for this gene is normal, but a child born with two defective alleles lives only a few years before fading into profound retardation and death. From a small blood sample it is easy to tell whether a potential parent has one or two normal alleles for this protein: like the pink and red snapdragons, one or two functional alleles have different levels of activity, both compatible with normal nerve cell development. While a carrier with one normal allele married to a person with two normal alleles can be sure that all of his or her children will be free of the disease, two carriers run a one-in-four risk of creating a doomed and tragic little life. For the past thirty years, potential parents from families at risk have been encouraged to take a test before conceiving a child. Couples who find they are both carriers of a defective Tay-Sachs allele may choose to have the cells of a fetus tested as well, and if it lacks the critical enzyme, to have it aborted.

Tay-Sachs disease can occur in anyone, but it appears with a very high frequency among the children of Europeans of Jewish ancestry.* Parental testing and counseling, and prenatal

---

* Like African Americans whose ancestors were made somewhat resistant to endemic malaria by one sickle-cell allele, many American Jews of European descent inherit the Tay-Sachs gene today because it gave some advantage to their ancestors. According to the anthropologist Jared Diamond, one Tay-Sachs allele — but not two — conferred some resistance to tuberculosis, a plague endemic to the ghettos of Europe.

testing and abortion, have together dropped the incidence of children born with Tay-Sachs in the American Jewish community by more than tenfold. In one famous example of sensitive genetic counseling, eligible young members of an Orthodox Jewish community were at great risk of having Tay-Sachs children but were opposed to abortion and in any case would not subject themselves to testing for fear that it would ruin their chances of marriage. Their predicament was solved by their creative rabbi. Himself a carrier of a mutant allele for the Tay-Sachs gene, the rabbi arranged a matchmaking service that would test its subscribers under conditions of total confidentiality, revealing the results only to him. Then, once a match was arranged, he would tell the bride and groom to consult a genetic counselor only if both carried a defective allele. Otherwise, he said nothing. In this way, two people learned about a potentially disastrous situation in private, when it could do no harm and only when it was necessary for them to know. If they then chose not to marry, only the rabbi would know the reason. From 1983 to 1987, more than four thousand young people were secretly tested in this way; six couples discreetly backed out of marriages that would have put them at risk of having a Tay-Sachs child, and not one child with Tay-Sachs was born to the community. This sort of sensitivity to the needs and fears of potential parents will be hard to match as the technology for detection of carriers of various inherited diseases improves, and genetic counseling will become more difficult to accomplish without imposing unexpected choices.

The alleles that cause PKU and Tay-Sachs need not be sought after as DNA sequences in chromosomes, because they signal themselves reliably in the bodies of their carriers. Most diseases, though, are not so flashy. We suspect they are inherited because they run in families, but in the absence of a chemical marker like the missing enzymes of PKU and Tay-Sachs, the complex phenotypes of such diseases do not shed light on the genomic problem. Cystic fibrosis is the most common inherited disease among Americans of European descent. About fifteen million Americans carry one allele for cystic fibrosis; among European Americans, the disease occurs in about one in sixteen hundred births. The symptoms are complex, including difficulty in breathing because of a thick mucus that fills the lungs. Current treatments prolong life and ease the pain, but they do not cure this disease. One ancient diagnostic sign prevails in all victims of cystic fibrosis, even infants: their sweat is excessively salty, suggesting that the disease may be the consequence of damage in some part of the cellular machinery responsible for moving salts in and out of cells.

The DNA difference responsible for the most common cystic fibrosis allele was recently discovered through the direct analysis of the genomes of carriers and victims. A team of Canadian and American physicians and geneticists succeeded in isolating a large, new gene from a chromosome band that had been associated with cystic fibrosis. Once it was recovered from the DNA of a healthy person, the sequence of bases in its DNA was deciphered, and from that, its encoded protein was predicted. Then the sequence of DNA in the same gene from patients with cystic fibrosis was deciphered. The DNA of more than 70 percent of these patients differed from normal in exactly the same way, showing a loss of three base pairs in the same place along the gene. With the normal and mutant alleles of this gene in hand, scientists have been quick to find, name, and study the normal and mutant versions of the protein it encodes. The protein normally sits in the membranes of the cells that line the airways and intestinal tract, regulating the amount of salt — or, more precisely, chloride ions — that can leave the cells. The cystic fibrosis mutant fails to regulate this flow properly, and the symptoms of cystic fibrosis ensue from this defect. This protein, called cystic fibrosis transmembrane conductance regulator or CFTR, had not been known to exist until this work.

Despite these discoveries, the treatment for cystic fibrosis still cannot go beyond the temporary amelioration of symptoms. In the near future we should see a new class of drugs that treat the cause, not the consequence, of this mutation; for example, drugs capable of restoring CFTR activity, if not CFTR itself, in the cells of cystic fibrosis victims. Also, in short order, tests for mutant CFTR activity carried out on cells and fetuses from potential carriers should become straightforward.

Such tests would give the much larger number of Americans who carry the cystic fibrosis mutant allele the same early warning available to carriers of the Tay-Sachs mutant allele.

The CFTR story is an early example of what has been called reverse genetics: the discovery first of the DNA difference associated with an inherited disease; then, by the comparative analysis of normal and mutant DNA, discovering the gene responsible for the disease and elucidating its normal role in healthy people. The allelic error of a disease turns out to be a good hook with which to fish out hitherto unknown human genes, and we can expect to see many more genes isolated, sequenced, and understood through reverse genetics.

∼

The second road from the analysis of the human genome to medical practice also presents difficult choices. The road begins with the technique of fertilization *in vitro* (from the Latin for "in glass"), or IVF. Until about twenty years ago, there were only two ways for an infertile couple to raise a child: adoption or — if the man's sperm cells were not viable — insemination with another man's semen. In either case, the child was still conceived and carried in the uterus. Then a burst of implausible but basically straightforward technology enabled physicians to stimulate a woman's ovary to produce egg cells that could be harvested from her body and mixed with a man's sperm in a glass dish. As soon as the first sperm deposited its nucleus inside an egg cell in one of these dishes, *in vitro* fertilization was complete; a new human genome had been launched into the world from the oddest of harbors. To ensure that this embryo had a chance to develop, it was then kept under the microscope for a few hours or days before being placed in the sort of environment every embryo its age takes for granted, a woman's uterus.

IVF is not a molecular technique, but the full panoply of analytical machinery based on DNA can be brought to bear on the naked, glass-enclosed, microscopically observed embryo as it goes through its first divisions. The same scans for aberrant alleles that work on one cell taken from the million billion cells of an adult will also work on one of the eight cells of an embryo whose entire existence began a few hours earlier in a glass dish. Although this point has an obvious logic, it may be harder to comprehend that an embryo can surrender one eighth of its substance without ill effect. But a very young embryo can afford to sacrifice one cell with no effect on its later development. Identical twins, for example, each grow up from one of the first two cells of a fertilized egg without any of the material, genetic or otherwise, in the other cell.

The technology for safely removing one cell from an eight-cell embryo is now well established: after several years of testing and use in animal husbandry, the technique was adapted for human embryos and successfully applied in 1991. If a test of the embryonic DNA finds at least one functional allele for the gene at risk, the remaining cells are watched for a while and then implanted to become a baby. If the DNA test shows that the alleles in the seven remaining cells lack, for example, even one normal allele for the CFTR or Tay-Sachs gene, they are forfeited — not aborted so much as never allowed to get fully under way.

Early detection and IVF are both becoming more common, but neither is becoming any simpler in its consequences. Will IVF technology move beyond its accepted role in countering infertility to become widely used by couples who are, or think they are, at some genetic risk? Each time a new gene is discovered, prenatal genetic testing looms larger as an issue; more choices open up for everyone, but they are not necessarily choices anyone would want. Who has the right to know the result of a prenatal DNA test: the mother, the father, the sperm donor, the egg donor, the bearer of the IVF fetus, the State, or — when they are all different — all of them? Who decides which alleles, and which embryos, escape abortion or qualify for reimplantation after IVF and DNA testing? As the technology for recovering mutant genes by reverse genetics improves, more and more women and men will face one or another Hobson's choice.

Also troubling is the high cost of these new tests and treatments. Many of these choices, difficult though they may be, are simply not available to Americans who lack medical insurance or even a job. In the United States today, more than forty million people are without any form of health insurance. Among them are likely to be perhaps a million carriers of a mutant cystic fibrosis allele. Without an agreement among the president, the Congress, and the nation's physicians to change the way medical care is provided, these citizens will never be able to afford a test for the mutant CFTR allele, let alone be given the guidance and facilities to act on the basis of its result. In the absence of insurance to pay for CFTR testing, IVF, or the eight-cell embryo test, they will become the parents of tens of thousands of carriers and thousands of children with cystic fibrosis each year, few of whom are likely to have access to any future gene-based tests or therapies to reverse their disease's fatal course.

These difficult issues seem benign when compared to those raised by a third road from the genetics laboratory to the world beyond, the one that goes through eugenics. Soon after Mendel's work was rediscovered a century ago, politicians as well as physicians saw uses for the knowledge of recessive alleles; inevitably, perhaps, patriots in several countries envisioned the possibilities inherent in a rational plan to improve the genetic quality of a nation. This third road has been little traveled lately, but in the first part of this century, the eugenics movement brought together some of the best geneticists and physicians and the worst chauvinists in the Western world. It was (and for some people still is) easy to endorse their early agenda: civilized people have an obligation to minimize the number of defective alleles in their chromosomes and in those of their descendants, replacing them with good, better, and best alleles.

Some eugenicists, however, were impatient with simple testing and counseling. Would it not be easier to cultivate the best selection of human alleles, they asked, if the wasteful, genetically risky business of having children were put under rational control, and easier still if the results of genetic analysis were fed into a state apparatus that would decide who could be born and who not? Germany was the most hospitable to the eugenics movement in the 1920s and 1930s. As they thought of ways to accomplish the "weeding and seeding" of human alleles, German eugenicists were first assisted, then taken over, by a political movement, a government, and a leader all driven by the crudest and most naive notions of national and racial purity. In that time and place it was only a short walk for many physicians, and for some professors of psychiatry, anthropology, zoology, and genetics, to go from theories of eugenics to the practice of mass murder.

Their downward spiral can be reconstructed from their writings and from the grim record they left behind in other ways. It went from an appreciation of the ability of recessive phenotypes to reappear unexpectedly after generations of silence to the clinical observation that certain mental diseases and physical deformities were inherited in this way; then to acquiescence in the nonsensical notion that some alleles observed national boundaries and religious distinctions; to the endorsement of the even more bizarre notion that within a country, a measurable set of alleles marked the national "type," so that persons whose phenotypes revealed their lack of these alleles could never be brought into the national fold; to the solemn decision that a life without such alleles was simply not worth living; to participation in the sterilization, and then the murder, of millions of people presumed — on the basis of such markers as the shape of their noses or the lilt to their voices — to lack these alleles in their chromosomes. When Adolf Hitler said "Politics is applied biology" in one of his most popular and successful slogans during Germany's fateful 1933 election campaign, he meant it.*

How could this happen? It is easy to see — standing on a mountain of ashes — where the scientists and doctors of Germany went off the deep end. But only twenty years before Hitler came to power, eugenics was a recognized, legitimate branch of genetics, and in Germany, the United States, and

---

* Benno Müller-Hill points out in his book *Murderous Science* that in March 1943 the internationally renowned human geneticist Professor Doctor Eugen Fischer, previously editor of the authoritative 1940 text *Human Heredity and Racial Hygiene*, wrote, "It is a rare and special good fortune for a theoretical science to flourish at a time when the prevailing ideology welcomes it, and its findings can immediately serve the policy of the state." A few months later the infamous Dr. Josef Mengele, then a young scientist at Fischer's Kaiser Wilhelm Institut in Berlin, was appointed the camp doctor in Auschwitz.

many Western countries it drew the attention of reasonable, educated people. Andrew Carnegie, for instance, was a generous and enthusiastic supporter of the international eugenics movement. He founded the Carnegie Station for Experimental Evolution at Cold Spring Harbor, Long Island, at the turn of the century. Charles Davenport, the director of the Cold Spring Harbor laboratory in the 1920s, contributed heavily to Congress's decisions in that decade to restrict immigration to the United States on "national" grounds. His testimony before Congress, and that of others, was full of eugenic contentions couched in the most scientific tone; for example, alcoholism, poverty, and avarice were argued to be "genes" inherited by people born of Irish, Italian, and Jewish parents, respectively.

The first wave of American eugenics was bad genetics, which caused a lot of suffering before it ran its course, but at least it stopped short of overriding our tradition that citizenship for immigrants and their children was a matter of law. The European eugenics movements of that period were not inhibited by such laws; in many countries they were given strength and legal standing by laws that inextricably linked full citizenship to notions of race and "blood." This coincidence of political and eugenic agendas helped eugenics in Germany to go off the tracks, derailed by an explosive combination of two mistakes. The first was the belief that an ideal human type exists. As a piece of science this makes little sense, flying as it does in the face of the first tenet of natural selection, that the survival of a species over the long term will depend above all on the existence of a maximum of variation from individual to individual. However, the notion took hold, and from it came the German eugenicists' notion of *Ballastexistenzen*, or "lives not worth living." In the years between Hitler's rise to power and the beginning of World War II, hundreds of thousands of Germans hospitalized with various genetic and mental ailments, others afflicted with alcoholism and the like, and still others with no particular problem except that they were attracted to people of the same sex were sterilized without their knowledge or acquiescence but with the agreement of their doctors. With the invasion of Poland in 1939, sterilization was succeeded by wartime euthanasia; these Germans died in hospitals and nursing homes by gas and lethal injection before the killing squads were vetted to new jobs in the concentration camps of the East.

The second error arose from the choice of phenotypes that would identify an individual whose appearance approached this ideal. In order for a program of controlled reproduction to be effective, all such ideal phenotypes had to breed true. The only phenotypes that are certain to breed true are those made by pairs of recessive alleles; dominant alleles cannot produce the surprise-free stability of phenotypes needed for a breeding agenda. Not surprisingly, then, German eugenicists planned to breed for the recessive phenotypes of tall height, blue eyes, straight blond hair, small ears, and a small nose. But choosing these as the desirable features of an ideal German also meant identifying the genetic, biological enemy. Each ideal phenotype could be overwhelmed at any time by a single unwanted but dominant allele that might come from a short, dark-eyed, curly-haired, large-eared, long-nosed wanderer. That was enough to ignite the interest of Hitler and anyone else short and dark who had notions of ethnically cleansing Germany of such people in order to build a "master race" of tall, blond, blue-eyed people.

Under Hitler the next step — marshaling the efforts of a nation behind a program of human breeding for recessive phenotypes — needed only one piece of scientifically meaningless, emotionally charged nonsense to throw the whole enterprise into malignant focus. This was the notion that despite all appearances to the contrary, every potential Jewish parent was inevitably the bearer of an undesirable, alien allele that would crush the ones Germany needed, the crazy idea that Jewishness was a single allele of a single gene. However inarticulately stated by Hitler's propagandists, this was the academically certified eugenic argument for the destruction by bullet, gas, and fire of German and then European Jewry, of Germans and others who had one Jewish grandparent, and especially of about a million Jewish children, some of them exactly my age.

∾

Eugenic programs need heading off early. Once established in the bureaucracy of a modern nation-state, any such agenda — especially one buttressed by the technological powers of modern biology — is likely to be able to survive war's defeat, dozens of elections, and decades of rejection by a multitude of governments. For example, almost sixty years after the Nürnberg Laws, the codification of "German-ness" as an inherited trait and the notion that the presence of "German" genes may be predicted from characteristics that are precisely only skin deep still inform contemporary German law. Passport laws automatically confer German citizenship on a class of people outside the national boundaries who are defined more or less as *eindeutschfähig*, "biologically eligible," to be German. All "non-Germans," on the other hand — including German-born, German-speaking people who do not fit this category — have a special set of questions to answer in applying for citizenship. Until 1991, one of these questions was, "What is the shape of your nose?"*

Nor did the egregious application of eugenics in the Third Reich vaccinate us against other pathological applications of biology to human affairs. Consider the common use of skin color as a marker of complicated, partly inherited, partly culturally modulated phenotypes, in particular the vastly complex and uniquely human traits of character and intelligence. This habit lives on even though there can be no impersonal, molecular shortcut to discovering a person's abilities. Indeed, many medical conditions — and most traits we dislike or qualities we admire — are not the products of single alleles, recessive or otherwise. To the extent that they are inherited at all, they are the consequence of the expression of large and unidentified assemblages of genes as well as of a lifetime of unpredictable interactions with other people. Even such a simple marker as an adult's height, for example, is determined in part by a set of about a hundred genes and in part by circumstance, upbringing, nutrition, and the like.

---

* In response to my request in 1992 for details on changes in German law regarding "non-Germans," the Ministry of the Interior in Berlin elaborated as follows: "Information on the shape of the face or nose may no longer be requested. However, unified forms for the new travel documents have not yet been introduced. Last year the federal government and the states agreed therefore, that until the design of the new travel documents had been decided upon, the old forms should be used. This decision was taken for reasons of thrift."

We can expect the genetic component of the human condition to become larger as we learn how to track alleles for numbers of genes at once. Aspects of sexual behavior, in particular, are likely to be mapped to the genome; after all, no part of behavior is likely to contribute more to the survival of a species than the will and the capacity to bear progeny. Dean Hamer of the NIH recently found, for instance, that some men displaying a common variant of sexual behavior in males — early-onset male homosexuality — inherit a specific small region of their mother's X chromosome. The search is on for the allele or alleles in this region of the X chromosome that contribute to male sexual behavior; the best bet is that these alleles will be different in heterosexual and homosexual men.

Every soldier surrenders his or her DNA to the military at recruitment, so that wartime casualties may be unambiguously identified. The policy of the U.S. armed forces toward homosexuals who wish to serve has recently become "Don't ask, don't tell, don't pursue." This policy, which now requires abstinence on the part of enlisted troops, will soon require abstinence of a different sort from the Pentagon: will the armed forces be able to avoid scanning millions of DNA samples for the appropriate sequences once a DNA associated with homosexual behavior becomes available? DNA differences as such cannot legitimately order humans in any hierarchy of present or future value, but it nevertheless seems likely that homosexual Americans will enjoy the same rights of privacy as do heterosexuals only if the president, the courts, and Congress agree in fairly short order that such DNA scans may not be carried out. Otherwise, homosexuals are likely to be the first Americans to become members of a new, DNA-based, genetic underclass.

Another lesson that must be drawn from this century's earlier, disastrous romance with applied eugenics is that we cannot possibly distill from the billions of evanescent drafts of the human genome a single, canonical text. "The human genome" does not exist except as an abstract notion, and while one or even a few alleles may one day be isolated and sequenced for every human gene, even this collection would be different in revealing and interesting ways from the particular human genome in you, or me, or anyone else who has ever lived or ever will live. Perhaps because they have to live with the shame of the Holocaust, postwar scientists and politicians in Germany have to this day shied away from concerted, large-scale work on the human genome, and German practices in genetic counseling are heavily tilted toward individual privacy. But in the United States — and surely eventually in Germany as well — future genetic counseling will inevitably provide the sort of information earlier eugenicists could only imagine.

About four hundred and fifty human diseases have already been linked to specific alleles of human genes, and there is every reason to suspect that mutations in many of the other hundred thousand or so human genes will also be associated with human disease in time. For instance, the special sort of mistake in DNA — extra repeats of a three-base sequence — that generates the Huntington's disease allele in one gene and the mental retardation of the Fragile X syndrome in another is also present in another hundred human genes, all of which wait to be analyzed and linked to their particular syndrome. The temptation to apply basic research on the human genome first to medicine, then to social policy, and then to traditionally private choices is sure to grow with time.

Neither great fame nor a track record of profound scientific insight is proof against this temptation. Consider this report of Sir Francis Crick's 1968 Godlee Lecture, taken from a news article in *Nature*, the same journal that fifteen years earlier had published his and Watson's discovery of DNA's structure and function:

If new biological advances demand a continuous readjustment of ethical ideas, how are people to be persuaded to adapt to the situation? Clearly by education, and Dr Crick did not think it right that religious instruction should be given to young children. Instead they should be taught the modern scientific view of man's place in the universe, in the world and in society, and the nature of scientific truth. Not only traditional religious views must be re-examined, but also what might be called liberal views about society. It is obvious that not all men are born equal and it is by no means clear that all races are equally gifted. . . . So important is it to understand the genetics of human endowment that parents should perhaps be permitted, Dr Crick said, to dedicate one of a pair of identical twins to society so that the two twins could be brought up in different environments and compared.

But do "new biological advances demand a continuous readjustment of ethical ideas"? The human genome is a text, but not a sacred one. As the six versions of one line from the Book of James show, even words held sacred by millions of people turn out to have many equally valid versions when examined closely; how much less likely is it that there will ever be a single, canonical human genome whose precise alleles we might hold up as perfect, sacred, or even special? Yet that assumption underlies a significant portion of current biomedical research and development. Each newly isolated and sequenced human gene invites the speculation that we have been brought closer to understanding what the ultimate, supremely "gifted" genome would be. Every time science gives us a new chance to dream this way, we are all obliged — as bearers of different but equally valid versions of the genomic book — to recognize the temptation, and to forswear it. Leaving the boundary between public and personal access to our genomes to the experts — biologists, physicians, lawyers, even Nobel laureates — will not do.

3

SENTENCES, SCULPTURES,

AND THE AMBIGUITIES

OF TRANSLATION

A COMPARISON OF DNA'S LANGUAGE with English helps put the genomic language and its meanings in their proper context. Like any other spoken language, English is a series of overlapping sounds issued in long streams carrying information, more like music than text; descriptions of the muscular movements giving rise to speech lie at the root of the English words "language" and "tongue." While we are born with the capacity to make about two hundred sounds, most languages use a duller spectrum of pops and whistles; spoken English, for instance, uses only about forty. Written English chops the flux of speech into separate sounds and captures them in a set of only twenty-six symbols. Using the sounds and twenty-six letters, we can speak and write all the hundreds of thousands of words in the English language and make up new ones at any time.*

Other languages follow the same pattern: whether spoken or written, they all assemble long strings of a rather small number of sounds or symbols; the various vowels and consonants of English, the consonant-vowel symbols of Korean, and the word symbols of Chinese are examples of how writing quan-

---

* Take the word "scientist," for instance. It did not exist until 1840, when the president of the Geological Society in London coined it to describe "a cultivator of science in general" and Charles Darwin in particular.

tizes spoken speech. Why are languages constructed this way? Why aren't they short agglomerations of ever larger numbers of different sounds or symbols? The rich informational content of a stretch of base pairs in a DNA molecule suggests the answer. The number of possible concatenations of a small number of symbols — or sounds — grows very rapidly with the length of the chain, whether it is a DNA sequence, a word, a sentence, or a song. For example, a speaker of English, using forty sounds, can assemble billions of "words" only six sounds in length.

Of course, almost all of these have never been said and never will be. Words are different from the vast excess of possible arbitrary strings of sounds or symbols in one critical way: each word has at least one meaning to the speaker. The concatenation of words affords great flexibility in the construction of different, unambiguous statements for different, unpredictable purposes. The principle of concatenation is so strong that human languages — and the genomic language — apply it twice. At the primary level, sounds or symbols are strung together to assemble a word, the minimal free form, the shortest natural figure of speech that can be spoken with meaning. Then words are strung together to make sentences; the possibilities for meaning in a sentence are greater than the sum of the meanings of its words. A language can be understood only by defining words by their functions in a sentence and studying the meanings of sentences.

The meanings of words can be classified according to any number of lexicons and dictionaries. Together, meanings define the conceptual system of a human language — the mental lexicon of that language — which resides in the minds of its speakers. Our mental lexicons seem constructed according to a basic rule that frees us from any limitation in how we choose to represent meaning. As the Swiss linguist Ferdinand de Saussure first pointed out in 1916, no sets of sounds or letters are restricted to a particular concept, nor is any concept naturally expressed only by one particular group of sounds or letters. Saussure gave the name "signifier" to the form of a word and "signified" to its concept. In his view, there can be no intrinsic link between the signified and the signifier, between the form of a word and its meaning. Despite the apparent arbitrariness of a language freed of any obligatory link between signifier and signified, languages do have a coherent structure, because both forms and meanings can be defined in terms of their relations with other forms and meanings, fusing signifier and signified in a set of "signs." Once these definitions are in place for a language, the mental lexicon stores them. The similarity of the lexicons in the minds of two people is what allows them to exchange meaningful conversation, and one person to comprehend what another has written.

The books people read and the words they speak may look and sound completely different as we go from one language to the next, but the attributes that define English as a language — letters concatenated into words and words into sentences; a syntax that defines meaningful sentences; a grammar for the parts of the sentence; a lexicon of meanings that carry over from one language to another; a later, archival, written version capturing the earlier spoken one — are shared by the other languages humans speak and write. Complex, complete meaning never resides at the level of sounds and symbols; it only begins at the secondary level of words.

The division between signifier and signified is deep and runs from philosophy to physiology. Fluency with signifiers can be measured by a capacity to spell words correctly; fluency with the signified concept is captured through a facility with analogies. Remarkably, lesions of the brain can affect one skill while leaving the other intact, suggesting that we keep signifier and signified separate even as we think. In philosophical terms, the lack of any direct, logical link between word and concept raises the stark obligation of the speaker and the hearer to confer agreed-upon meanings to the words they use. If audiences — you and I — cannot agree on a common set of signified meanings, our discourse must collapse into mere word play, a chaos from which words themselves cannot preserve us.

The deepest syntactical division in any language separates how words can be used in a sentence, distinguishing nouns from verbs. English syntax shares one universal rule with the syntax of all other spoken languages: every sentence has at least one noun part and one verb part. In English, a further syntactical rule puts the noun part before the verb part: the simplest grammatical English sentences take the form "noun verb," as in the laconic "He died." Beyond this, syntax provides hierarchical rules for the formation of more complex sentences; these rules — together with word endings that convey tense, number, and gender — make English highly complex and flexible. But all English sentences, like those in all other languages, are built from nested sets of one basic unit, a noun part with a verb part.*

Even though a sentence is more than the sum of its words, the concepts the words conventionally express must be part of the sentence's meaning. The semantics of a language define the meanings of its sentences in spite of the ambiguity and multiplicity of meanings carried by many words. For example, when the distinction between two meanings of "see" is intentionally blurred — "I see your point" — the power of the mixed use of the verb lies in its ability to convey the concrete meaning of seeing something with one's eyes along with the abstract notion of agreement. Context removes ambiguity from the meaning of a sentence, as it does for the words in a sentence.

Because words can have more than one meaning — this is called polysemy — sentences of great similarity may have completely different meanings. Polysemy is common in all languages, although more so in some (like Chinese) than others. My favorite example of the potential for confusion created by polysemy is a pair of sentences of the sort designed to stymie computers that try to read English by word comparison rather than meaning and syntax: "Time flies like an arrow" and "Fruit flies like bananas." In the first, "flies" is the verb; in the second it is a portion of the noun part. In the first, "like" is an element of the noun part, comparing the subject of the

---

* In English and other members of the Indo-European family of languages, gender is based on distinctions between the male and female sexes, so that nouns may be masculine, feminine, or neuter. But gender is a linguistic class of noun forms, not necessarily the same as sex. In Bantu, for instance, nouns fall into one of nineteen genders, including thin, human, female, animal, body part, and location.

sentence to another noun; in the second, it is a portion of the verb. The fact that languages permit polysemy is an independent confirmation of the distinction between signifier and signified and of the existence of signified meanings: in order for polysemy to be a problem, we must be storing polysemous words in our heads in more than one "place" according to their different signified meanings.

~

To understand the linguistic properties, syntax, grammar, and semantics of a human genome, we have to look at how a gene speaks to a cell and at the meanings of what it says. When we look inside the nucleus of one of our cells we see — after a moment to adjust for the necessary shift in scale — DNA in letters that connect to form words and words that connect to form sentences. We have already encountered these sentences: they are alleles, the specific versions of a gene. We know from genetics that alleles lie on a chromosome one after another in no particular order; they rarely form paragraphs or longer blocks of meaningful text when transliterated in the order we find them on a chromosome. The genome is not a book, composed to be read from beginning to end, but a lexicon, a collection of arbitrarily ordered sentences, similar to the arbitrary alphabetical order of entries in an encyclopedia.

The written lexicon of DNA that fills the nucleus of every cell in our bodies will remain silent until each cell uses it by excerpting its own set of transcribed and edited quotations called messenger RNAs. Messenger RNAs are signifiers as well, evanescent strings of gene domains. The cell shows its interpretation of their underlying meanings by converting them into proteins, which behave like portable, three-dimensional signified concepts. While every gene is a double helix no matter what its information, different proteins have different shapes, and each shape confers a capacity to carry out the actions described in its encoding gene. This leap from DNA to protein, from line to shape, from signifier to signified, makes a spectacular difference between linguistic and genomic texts: it is as if each sentence in an encyclopedia were folded into a unique origami sculpture that carried its meaning in its shape.

By translating enough texts into proteins with various shapes and consequent functions, a cell draws from its genome the capacity to carry out the multiplicity of interactive chemical changes that allow us to call it a living thing.

Bent, knotted, and folded, a cell's protein sculptures look like the animals in Alexander Calder's wire circus. They are made of thin molecular chains — thinner even than DNA's double helix — whose twenty links are different from one another and as bumpy as the wire that carries Christmas lights around a tree. These links are called amino acids. Like the four bases on a strand of DNA, the twenty amino acids can be hooked together in any order; they can appear at different places on the chain any number of times; and, like the four base pairs, each of the amino acids has a different molecular shape. But nothing prevents each sequence of amino acids from folding into three dimensions. In fact, the sequence of amino acids in a protein completely determines its final shape as it folds up on itself in its own specific way.

The specificity of protein folding is startling: billions of copies of a single protein can be purified from the cell and merged together in perfect alignment as a crystal. Such crystallization would be impossible if two copies of the same protein folded up even slightly differently. Each meaning of a sentence in the written language of DNA is thus manifested in a protein, because the sequence of DNA base pairs determines the sequence of amino acids in a protein, and the sequence of amino acids determines the three-dimensional shape into which the protein will fold.

Just as we can transliterate a sequence of base pairs in DNA from the full complexity of its three-dimensional, molecular shape into a sequence of four letters, we can convert the sequence of amino acids making up a protein into a string of twenty English letters. In this transliteration, A is the amino acid alanine instead of the base adenine, C is cysteine instead of cytosine, D is aspartic acid, and so forth, ending with Y for tyrosine. Though some of the letters are the same for both DNA and protein sequences, the twenty possible amino acids available for each position in a protein's chain mean that an amino acid sequence of any length can have many more possible meanings than a DNA sequence of the same length. For example, in DNA the sequence CGAT — cytosine-guanine-adenine-thymine — is one of 256 possible DNA sequences that can be constructed from four base pairs, whereas the same sequence in a protein — cysteine-glycine-alanine-threonine — is one of 160,000 possible stretches of four amino acids.

As with a DNA sequence, the letters of an encoded amino acid sequence are signifiers; only the three-dimensional shape of a protein carries the meaning of its encoding DNA. The freedom to fold and twist into three-dimensional structures gives a protein its capacity to express the meaning of the DNA that encoded it. Any two proteins can be very different in overall shape, and significant differences can occur with only very small differences in amino acid sequence. Hemoglobin, for example, is an amino acid chain with 141 links; a switch of only one amino acid in the chain distinguishes normal hemoglobin from its sickle-cell variant, yet the two proteins fold into shapes so different that a red blood cell containing the variant is distorted in shape.

What the folding chain of amino acids does so precisely, it does by rules we do not yet understand. Schrödinger's oxymoron about DNA was resolved when we understood how a double helix could be, in fact, an aperiodic crystal. The proteins encoded by the aperiodic crystal are also oxymoronic: their individual shapes are precisely unpredictable. So long as this is true, the genomic language, like our own languages, will not have a logical link between signifier and signified. This will not prevent its being read or understood; rather, it will assure that DNA remains a language expressing as full a range of meanings through arbitrary signifiers as any other language.

~

The DNA language of the cell, and the way it is made manifest in protein, find parallels in the Greek, cursive Egyptian, and hieroglyphic Egyptian inscriptions found on the Rosetta Stone. Unearthed in 1799, the stone had been inscribed by the priests of Memphis about eight hundred years earlier, when Greek — the dominant language of coastal Egypt — was used even by priests of the ancient Egyptian religion. The stone had three texts in horizontal bands: fourteen lines of hieroglyphic bas-reliefs on the top, thirty-two lines of incised cursive, demotic Egyptian text in the middle, and fifty-four lines of the Greek of the day at the bottom. The Greek stated that the document set forth the same text concerning royal matters in three scripts. Nevertheless, its translation from Greek to the other two scripts was problematic. Hieroglyphs seemed too subtle and mysterious for simple translation. The demotic and Greek scripts were understood to encompass words by a similar use of symbols for sounds, but ever since Pythagoras, each hieroglyphic symbol was thought to be a complete, allegorical message, and the discoverers knew no way to translate between letters and allegorical symbols.

Fifteen years after it had been discovered and taken to Europe, the Rosetta Stone's translation was begun by Thomas Young, an English physician and physicist. Young saw that the hieroglyphs representing royalty were surrounded by a carved protective cord called a cartouche. He correctly guessed that at least within a cartouche, hieroglyphs had the values of sounds after all, in particular the consonant sounds of such royal names as Ptolemy and Cleopatra. Young was only partially successful in decoding these sounds, misattributing about as many sounds as he got right. The young French historian and linguist Jean-François Champollion — master of Latin, Greek, and six Oriental languages by the age of sixteen — broke the Rosetta Stone's code in 1821–22 by showing that hieroglyphic writing was a rebus. Some signs were alphabetic, as Young had inferred, but others were syllabic and still others were indeed symbolic, representing and summarizing a whole idea or object previously expressed alphabetically. In decoding the stone, Champollion also showed that it had been written first in Greek, then translated to hieroglyphics — that is, from alphabet to rebus, not the other way around. Although Greek was the prevailing language when the stone was carved, the full religious meaning of the text could only be found in its hieroglyphic representation, and the priests had taken care to preserve that meaning.

As with the Rosetta Stone, the last translation of a gene is the most complicated. DNA and the stone both carry a linear representation of a text into a sculptural one. In both, information is translated from an alphabetic sentence of many letters (base pairs or the Greek alphabet) to a second alphabetic language of letters (amino acids or the demotic Egyptian alphabet), then to a three-dimensional, sculptured figure (protein or hieroglyph). While we cannot yet unpack the meanings of a gene the way a protein does each time it folds into its native, active form, we have learned how a protein is put together from the information in a gene. This is called translation, but it is an inexact use of the word, since the process does not merely carry the meaning of a sentence from one alphabet to another — although it does do that — but also allows the string of letters in the second, protein alphabet to immediately fold itself up, thereby enacting each sentence's meaning.

If a protein is like a hieroglyph in the sense of being sculpted rather than written, the analogy ends there. A protein *is* the meaning of the DNA word, not just its translation into a pictographic language. Proteins can move about, and they are sent from the genome to a particular audience far from the chromosomes of the cell. Wherever it goes, a protein conveys the meaning of its gene, whether to other proteins, to DNA, or even to other cells, and it speaks in a language to which genes, cells, tissues, and organs all respond. The DNA genome of a cell is in constant conversation with itself through the translation of particular genes into DNA-binding proteins. Through other proteins, every genome is also in dialogue with other regions of its cell beyond the nucleus and with genomes of other cells of the body. The interplay between genes and their proteins in the genome can be as simple as the operation of a thermostat or as complex but orderly as the proceedings of a courtroom. Genes make proteins, proteins make genes; together they make a cell: a chemical dialectic makes us all, because a gene speaks only when it is spoken to.

The cell's language is grammatical, with a simple syntactical structure. Genes — the stretches of DNA that are capable of speaking through the proteins they encode — are divided into words, called domains, which are quite like verbs and nouns. Just as an English sentence must have a noun and a verb, so proteins must have at least two functional parts. Verb domains ("do this") convey the specific action a protein will take, while noun domains ("to that") convey the target of the protein's action. The minimal protein domains for "do this" and "to that" may be given greater specificity — or more subtle meanings — by other protein domains. A gene's domains are laid out in a row in its DNA as contiguous sequences of base pairs. But because a protein is capable of folding into three dimensions, its domains need not be formed out of one continuous run of amino acids. A protein can fold over itself, bringing together the front and back ends of the amino acid chain into a single, crisscrossed basket of meaning. Grammar is preserved despite the non-linear connection of genes to their protein domains. The part of a gene that is translated will yield at least one verb domain and one noun domain within the protein, and the positions of these domains, like the order of words in a Latin sentence, are less important than the precise form of each.

Proteins and their domains work by recognizing the three-dimensional shapes of other molecules. The proteins called enzymes change the molecules that fit into them. The outer surface of an enzyme is indented with pockets of various sizes that fit with remarkable exactitude around another molecule. What happens after the instant of recognition by touch depends on the rest of the enzyme's structure. Some enzymes break a bond between atoms in the molecule they bind to and then let the broken pieces go. These enzymes are the digestive apparatus of a cell, reducing large molecules to small ones. Other enzymes accomplish the reverse, finding two smaller molecules and linking them by abetting the formation of a specific chemical bond between them; the DNA polymerase that creates two molecules of DNA from the separated strands of one DNA is such an enzyme. But not all proteins are enzymes; indeed, the most common protein in our bodies is collagen, which is more commonly known as the primary material of Jell-O. Collagen accumulates in the spaces between cells. One thread of collagen recognizes another in such a way that a multiplicity of threads line up, coil around one another, and weave a thick mat around the cells, holding the entire body together.

Our immunity from infection depends entirely upon the

ability of proteins to recognize one another. All living things, ourselves included, are no more than fertilizer beds for other, smaller organisms. These invisible creatures — viruses, bacteria, protozoa, yeasts, and molds — can quickly and completely break down the cells of our bodies, using us as food just as we use dead plants and animals to feed ourselves. While we are alive, our blood wards off these invaders with a complicated set of cells and secreted proteins, the immune system. Cells in our blood can rapidly and efficiently engulf and kill any invading microorganism, providing we can tag the invader with another set of proteins called immunoglobulins or antibodies. We carry millions of immunoglobulins in our blood; there are very few foreign organisms whose surface molecules cannot be recognized by at least some immunoglobulins.*

Each of the immunoglobulins in our blood has one or more regions, called antigen-binding sites, which can recognize a single patch about twenty by twenty atoms in area on the surface of another molecule. Further, each immunoglobulin is always testing the blood for foreigners from its perch on the outer membrane of a few immune system cells. When a particular cellbound immunoglobulin recognizes an invading molecule, the cell producing that immunoglobulin is stimulated to divide, generating a clone of identical cells that make precisely the right immunoglobulin to bind to the offending molecule. We can feel this happening during a viral infection when we notice we have "swollen glands." These are nodes of immune system cells, growing and secreting antibodies.

In the immune system, as everywhere else in the body, the protein-encoding domains of a gene are silent until they are translated, like the unspoken words of a written sentence. Or think of an orchestra: when it performs — no matter how

---

* How do we manage to make millions of different immunoglobulins with a genome that has no more than about a hundred thousand different genes altogether? Think of the children's book that has a set of drawings of fanciful people and animals, with each page slit horizontally in thirds. By combining the top of one drawing with the middle of a second and the bottom of a third drawing, one can assemble a vast assortment of amusing recombinant portraits, many more than could be printed and bound in one book. Our bodies follow this strategy to distribute millions of reassembled immunoglobulin genes into cells of the immune system so that each new gene is in at least a few immune cells.

atonal the composition — most of the notes from most of the instruments are not played most of the time. If all the notes were played all the time, the result would be not music but noise. So it is with the genome: every cell of the body has the same full set of genes capable of ordering the construction of an entire body, but most of the time, most genes in most cells are silent.

Before a gene can be translated, a set of proteins must act on its regulatory region, a stretch of DNA next to the gene itself. Binding to this silent DNA, a paragraph's worth of regulatory proteins — which are themselves made from information in the DNA of the cell — together determine whether a gene's single sentence will be translated. Some of these proteins are tissue specific, others respond to environmental or hormonal signals: certain plant genes, for example, are sensitive to sunlight. Shaded endive is white rather than green because its pigment genes have been regulated to keep silent in the dark.

The regulatory region of a gene has a second syntactical structure of its own, one that is not limited by the requirements of communication at a distance through translation. While the spoken, translated portion of a gene is syntactically of the form "do this to that," the unspoken regulatory portion is a different sort of command: "Now, here, begin translation." The syntactical rule for the regulatory portion of a gene sentence is that the domain for "here" must follow the domain for "now" and immediately precede the first domain of the translated portion of the gene. Following a rule reminiscent of the way an English sentence must have the noun before the verb, the untranslated command of a gene sentence always precedes the translated portion, so that the complete written DNA sentence we call a gene will have the regular form "Now, here, begin translation: do this to that."

The domain for "here" is typically the same from gene to gene, and because it includes a string of alternating A-T and T-A base pairs, the domain is called the TATA box. But the "now" domain that begins a gene's sentence is notoriously complicated and different for every gene. In fact the "now" domain is often really a group of domains called response elements, and a gene with a large number of response elements that together mean "now" will be translated only in response to a set of regulatory proteins. Some genes that are active only in liver cells, for example, have at least five different "now" domains. While cells of many tissues make one or more of the regulatory proteins needed to activate a liver-specific gene, only liver cells contain all the proteins necessary to signal that translation should occur from these genes.

Some DNA sequences can be recognized only by a complex of two or more regulatory proteins, just as some locks may need two keys to be opened or some checks two signatures to be cashed. One way two regulatory proteins can join and together bind to a response element involves a protein domain called the leucine zipper, a long coil of amino acids with a particularly oily one, leucine, at every seventh place. When the coiled leucine zipper domains of two proteins meet, their leucines interdigitate like the teeth of a zipper, each supplying the other's leucines with a comfortably oily environment. Zipped together, two leucine zipper molecules can gain the ability to bind to a jointly recognized response element. Domains like the leucine zipper allow for combinations of regulatory proteins to address a large number of specific regulatory sequences.

Without a score, an orchestra must be silent or risk disharmony. But even with a completely annotated and fully rehearsed composition, no orchestra plays music unless all of the musicians are prepared to be silent much of the time, responding only when the score calls for their contribution. In the same way, the machinery of the cell is silent until the genome conducts an orderly dialogue between regulatory domains and DNA-binding proteins. That exchange determines which genes get translated at what moment in a given tissue; the resulting cascades of activation — especially of the genes coding for regulatory proteins themselves — can order the formation of a complex tissue. Symphony or embryo, the principle is the same: the more complex the pattern, the more important the silences.

~

Once regulatory proteins bind to the appropriate region of a gene, it is ready for translation. Like the hands of a pianist playing one complicated chord again and again, the proteins play the "now and here" chord, thus permitting the gene to express itself. As that chord plays, a disposable copy of the information needed to complete the gene's sentence is sent from the gene's base sequence to the cytoplasm, where the machinery of translation sits. Making this disposable copy is called transcription, and the copy is assembled out of RNA, ribonucleic acid, an ancient cousin of DNA. RNA is usually found in the cell as a single strand rather than a double helix, although a strand of RNA with the proper sequence of complementary bases can fold back on itself to form a short hairpin of double helix.* RNA transcripts are copied from a DNA template by an enzyme called RNA polymerase. As it begins to travel along the gene, RNA polymerase twists the DNA so that its two strands are slightly unwound, separating a dozen or so bases on each strand from their paired bases on the other. The enzyme then brings complementary RNA bases to one of the two naked DNA strands and links them to create a new RNA, whose base sequence is complementary to one DNA strand. For instance, whenever the RNA polymerase detects a G on the DNA strand, it grabs a C from the soup of molecules swimming in the vicinity and pairs it to the G. Every T, likewise, gets an A. But when there is an A in the DNA sequence, the paired base is not T but a close analog called uridine (U). So, for example, if the template DNA strand had the sequence GCAT>, its transcript would have the sequence <CGUA rather than <CGTA. The RNA polymerase continues down the gene, stitching together this complementary RNA transcript, until it reaches a short sequence that signals it to stop. At this DNA domain the RNA polymerase lets go of both DNA and the transcript, and the DNA rewinds itself into a double helix.

The notion of a polymerase grabbing on to a DNA strand is more than a metaphor. Two relatives of RNA polymerase — DNA polymerase and the reverse transcriptase (RT) enzyme from the HIV virus — attach to DNA the way a sailor's hand catches a rope: the domains of these proteins both take the

---

* Some RNAs take on sculptural shapes by the interaction of hairpin loops. These folded RNAs can be recognized by specific proteins and assembled into complex RNA-protein structures.

precise form of a person's right hand. The structures of these two polymerases confirm that the order of domain words is not critical to a gene's meaning. The DNA polymerase gene encodes the domains in the order thumb-palm-fingers-palm, whereas the RT gene has its hand in the domain order: finger-palm-finger-palm-thumb. But because similar meanings are expressed by similar forms, these two proteins both fold into tiny, hand-shaped sculptures.

Before translation can begin in the cytoplasm, the RNA transcript of a gene must be edited and bound within the nucleus. The editing is accomplished by a set of RNA-protein complexes called sNRPs (pronounced *snurps*, short for small nuclear ribonucleo-proteins). These sNRPs bind to particular stretches of RNA in a transcript called introns, clip them out, and discard them. Once sNRPs reconnect the free ends of the remaining transcript, the edited strand of RNA no longer carries a base sequence identical to its gene, or to any other gene. Editing thus changes a transcript profoundly, deleting a portion of the text as a newspaper editor would to make a headline fit. And since introns can be spliced out in various ways from a single transcript, editing introns is a source of polysemy in DNA sentences as well, multiplying the final meanings of a gene before translation by unpacking two or more edited versions of a transcript from a single gene's DNA sequence. To protect edited transcripts from getting dog-eared, nuclear proteins add a chemical cap to one end of each and a tail — a stretch of As — to the other. Finally, the clipped, coiffed transcripts — now messenger RNAs — are ready to be sent to the cytoplasm to be translated.

Just as cells containing different mixes of regulatory proteins will open different genes, cells with different mixes of sNRPs may edit a transcript from the same gene differently, thereby making different proteins from the same opened gene. In just this way, for example, the gene for the muscle protein tropomyosin provides a panoply of proteins to different cells of the body. Tropomyosins regulate the speed and strength of muscle contraction. In some tissues, like the light meat of a chicken, the muscle must contract forcefully and quickly. In other tissues — the dark meat — the muscle contracts more slowly. In still other tissues, like the smooth muscle lining arteries, muscle cells are subject to hormonal regulation. Finally, fibroblasts, the cells that secrete collagen, must migrate to fill a wound. These cells too have a form of tropomyosin that is different from all the others.

The tropomyosin gene has a set of thirteen protein domains, each separated by an intron. The ability to make a particular set of splices in the transcripts of genes like tropomyosin allows us to make different kinds of muscle cells in different parts of the body. By alternative splicing in different tissues, any one of at least seven messenger RNAs is made in one or another tissue, generating different versions of tropomyosin, each conveying a slightly different meaning in its own cellular context. Imagine a conversation among various regulatory proteins and sNRPs concerned with muscle development as a chick is developing in its egg. Converting the tropomyosin gene's information into English, where "Now, here," is the gene's regulatory sequence, the gene would be read as "Now, here, begin translation: Make [intron] some [intron] light [intron] dark [intron] gizzard [intron] meat." When regulatory proteins bind to "Now, here," RNA polymerase begins transcription of that sentence. In breast muscle, editing the introns would generate a messenger with the intron-free instruction "Make some light meat," while leg muscle would edit the same transcript into "Make some dark meat." Both tissue-specific readings — and the other variants in gizzards and the like — would conserve the domains for "make," "some," and "meat."

～

The common meanings of transcription and translation imply a certain rigidity in the former and the possibility of creative variation in the latter. Just the reverse is true for the molecular events we give these labels to. After an immensely complex series of interactions among many proteins, each carrying the meaning of its own gene, a cell decides when and where to transcribe a gene and how to splice the transcript, and different cells make different decisions. But once a cell does make and edit a transcript into a messenger RNA, translation into pro-

tein proceeds at once. A messenger RNA, like a grooved phonograph record, is the portable representation of the base-pair
bumps of its gene, and the translating machinery of the cell,
like a juke box, will play the music of any record the cell
brings to it. It will convert the messenger RNA's linear text,
its sequence of base bumps, into a chain of amino acids. These
will then become the meaning of the gene as they fold up into
a fully articulate, movable, active protein.

To put the messenger RNAs created by transcription and
editing to work, proteins shuttling between nucleus and cytoplasm carry them by cap and tail through doorlike pores out
of the nucleus into the cytoplasm. The cytoplasm lies between
the nucleus and the cell's outer membrane; to return to the
comparison between a cell and old Jerusalem, the cytoplasm is
the part of the Old City outside the Temple but inside the
walls. There messenger RNAs are fed into the machinery that
will convert their sequence of bases into a chain of amino
acids. Like everything else that is big and complicated in a cell,
this machinery is built from prefabricated modules, and the
translation modules are another large, complex collection of
RNAs and proteins, called ribosomes.

The cytoplasm's assembly line of ribosomes takes hold of
the front end of a messenger RNA and uses base pairing once
again, this time to bring amino acids together in the order
directed by the messenger. By themselves, amino acids cannot
enter this assembly line, because they are the wrong shape for
base pairing. To prepare for their concatenation into a protein,
each of the twenty amino acids is first attached to its own
special small RNA, called a transfer RNA. Each transfer RNA
has a run of three bases that can form base pairs with three
complementary bases on the messenger RNA. Accuracy in
translation is assured because only one amino acid can be
connected to a given transfer RNA; moreover, the correspond-
ing set of three complementary bases on the messenger RNA
encodes only that particular amino acid. The assembly line
uses this method of base pairing to align and link a string of
amino acids that matches the sequence of bases of a messenger
RNA.

Transfer RNAs embody the act of translation: each is bilin-
gual and unambiguous. They function a lot like a Chinese

chop block, a handy carved stone about the size and shape of a knife handle that combines the functions of seal, stamp, and signature. Like the name of a person incised in Chinese ideograms on the bottom of the chop block, an amino acid sits at one end of a transfer RNA. The set of RNA bases that form base pairs with the messenger RNA at the other end of a transfer RNA is like a label on the top of the block with the same name in English letters; a person who reads English but not Chinese could string together a row of these blocks to print a list that would be clear to a person who read Chinese.

The cytoplasm of every cell has many copies of each amino acid as well as many copies of sixty-one kinds of transfer RNAs. Though small, each transfer RNA is flexible enough to fold back on itself into three short base-paired stems and three RNA loops; the flattened, two-dimensional path of the bases in a transfer RNA looks like a cartoon drawing of Mickey Mouse's glove or a three-leaf clover. One end of each transfer RNA — the stem of the clover — is a site for the attachment of an amino acid. An enzyme recognizes each transfer RNA and attaches a specific amino acid to its stem. Each kind of transfer RNA has a different sequence of three RNA bases at the outer edge of one of its loops; the base sequence of this triplet predicts with certainty which amino acid will be attached to a given transfer RNA's stem.

The correspondence between the diagnostic triplet in a transfer RNA and the amino acid bound elsewhere permits the ribosomal machinery to translate accurately from base sequence to amino acid sequence. There are sixty-four possible ways to make messenger RNA triplets by connecting three RNA bases in a row, by choosing one of four (A, G, C, or U) for the first base, one of four for the second, and one of four for the third. All but three triplets are complementary to one or another transfer RNA's diagnostic triplet. In the ribosomal machine, each of the coding sixty-one triplets will base pair with a particular transfer RNA by its diagnostic triplet and thereby recognize a particular amino acid. The sixty-one messenger RNA triplets coding for amino acids in this fashion are called codons, and the table of correspondences between codons and amino acids is called the genetic code.

The translation of a gene sentence into protein according to this code — a pretty much universal, four-billion-year-old molecular Esperanto — begins as the transfer RNA–amino acid hybrids and the messenger RNA enter the ribosome assembly line. A ribosome brings two transfer RNAs, with their amino acids in place, into base-pairing contact with two successive triplets of messenger RNA bases. Gripped by the ribosome, the amino acids attached to the two transfer RNAs find themselves so close together that the chemical bond holding the first amino acid to its transfer RNA is broken, at which point it attaches instead to the second amino acid, which remains attached to its transfer RNA. With the transfer of a chemical bond, the first transfer RNA is no longer connected to an amino acid, and it no longer fits in the ribosome. It immediately falls out, displaced by the second transfer RNA, the tail of which now carries not one but two amino acids. The ribosome then ratchets the messenger RNA along by precisely three bases, the way the pawls of an escape mechanism advance a clock's gears.

As the ribosome travels in three-base jumps along the messenger RNA, it holds on to the transfer RNA attached to the growing protein's most recently added amino acid. Each transfer RNA that comes into the ribosomal machinery adds its amino acid to the growing end of the protein and in turn becomes the temporary link holding the growing protein to the machinery. Each time the assembly line adds a new amino acid to the growing chain in this way, it pulls the messenger RNA three bases farther along. These jumps enable the ribosome to base pair the proper transfer RNA — with its attached amino acid — to each successive triplet along the messenger RNA, at each step adding the proper amino acid to a growing protein. The assembly line is a zipper with a difference: the ribosome unzips a succession of transfer RNAs from the messenger RNA, growing a new protein as it goes.

The ribosome cannot know where to end the translation; such punctuations are encoded in the messenger RNA by UGA>, UAA>, or UAG>, the remaining three of the sixty-four possible codons. The three stop codons, as they are called, do not bind any transfer RNA. Instead, each binds a cytoplasmic protein that dislodges the new amino acid chain from its transfer RNA. This disassembles the entire machine, freeing messenger RNA and ribosomes for another round of translation. Stop codons usually restrict protein coding to one strand of a gene's DNA. Even though in principle both strands of a gene might carry coding information, once one strand uses certain triplets to code for amino acids, the base-pairing rules make it likely that stop codons will be present on the other strand. For example, the stop codon UGA> will be present on a strand of DNA each time the other strand uses UCA> to encode the amino acid serine.

With sixty-one transfer RNAs and their complementary messenger RNA codons but only twenty amino acids, all but two of the twenty amino acids are able to find and attach to more than one transfer RNA. Three amino acids (serine, leucine, and arginine) are encoded by six codons; five amino acids (valine, alanine, glycine, proline, and threonine) are encoded by four codons; one (isoleucine) is encoded by three codons; and nine (phenylalanine, tyrosine, cysteine, lysine, histidine, glutamine, glutamic acid, asparagine, and aspartic acid) are encoded by two codons. Only methionine and tryptophan are encoded by unique codons. Francis Crick first pointed out the special nature of a codon's first two bases in assigning amino acid specificity: the alignment of transfer RNA and messenger RNA on the ribosome depends critically on pairing to the first two bases of a messenger RNA's triplet, permitting a certain degree of "wobble" between transfer RNA and messenger RNA at each third base. The redundancy of the genetic code, coupled with wobble, allows more than one transfer RNA to place the right amino acid in the right place along a messenger RNA–ribosome assembly line, speeding up translation.

The polysemy interjected by editing adds to an ambiguity already brought on by redundancy in the genetic code: two or more DNA sequences may in fact encode the same protein by using redundant codons to order the assembly of the same amino acid sequence. This ambiguity is one way: looking at a protein's amino acid sequence, it is not possible to predict the DNA sequence that encoded it. In contrast, every messenger sequence is unambiguous: it leads to only one protein. A one-way ambiguity of this sort is called a degeneracy, and we say that the genetic code is degenerate. This degeneracy enables a stretch of DNA to encode two different proteins on its two strands by allowing each strand to avoid the use of the three complementary codons that would bring a stop to the translation of a messenger made from the other strand.* In certain small genomes whose reading space is cramped, in particular those of some viruses, genes are indeed transcribed from both strands, and small regions of transcriptional overlap may be generated by DNA double entendre. Besides viruses, each of us also carries many copies of an overlapped, bidirectionally coding genome in the cytoplasm of each of our cells, in hundreds of energy-producing machines called mitochondria.

Because all living things use the genetic code, it must have originated very early in the history of life on Earth. Presumably the twenty amino acids were each essential to the development of cells, and the proteins of living things have been made of all twenty ever since. But it is easy to imagine that, originally, life may have been based on a much simpler genetic code. If the earliest forms of life could have existed with only fifteen instead of twenty amino acids, a minimal genetic code may have had fifteen codons for the amino acids and one for a stop. Sixteen codons can be encoded by the sixteen possible doublets of bases (four possibilities for the first base times four for the second), so such a code could have been, not only simpler, but unambiguous and nondegenerate as well; each gene would be only two thirds the length of genes today, a considerable saving of molecular materials over time. But the genetic code, like much else in life, is historical rather than rational, and once twenty amino acids worked, natural selec-

---

* The codon UCA>, for example, codes for serine. DNA encoding it on one strand must insert the complementary stop codon <AGU (or UGA>) in a messenger made from the complementary strand. The degeneracy of the code, however, allows the serine to be coded for by five other codons: UCU>, UCC>, UCG>, AGU>, or AGC>; these choices would put the codons for arginine, glycine, arginine, threonine, or alanine in the complementary messenger RNA rather than the stop codon UGA>. Similarly, UUA> and CUA> both code for leucine but would insert the complementary stop codons UAA> or UAG>; therefore DNAs that use both strands to code for different proteins must use the other four codons for leucine: UUG>, CUU>, CUC>, or CUG>.

tion did not have occasion to discard any of them. The degeneracy of the genetic code tells us that natural selection is profoundly conservative. It does not abandon old mechanisms easily, even in the process of assembling a rational human being who might imagine a more efficient translation system.

～

Mutations provide natural opportunities for interpretation, and we can use them to tease out some of the subtleties of a cell's reading of its DNA. For instance, the fact that codons are three base pairs long means that while any gain or loss of base pairs will be a mutation, the gain or loss of one or two base pairs may have a far greater effect on the gene's encoded protein than the gain or loss of three, six, or any multiple of three base pairs. The translating assembly line can accommodate the addition of an extra amino acid, or the loss of a few, and still turn out a respectable if queer version of a protein. These changes, called missense mutations, may generate a phenotype that is indistinguishable from normal or one that is clearly wrong from the start; there is no way to predict the consequences to a protein of losing a single amino acid. If a missense mutation damages or deletes a protein's "do this" domains, the mutant protein will be inactive. If the "to that" domains are hit, an active but poorly focused protein, worse than none in some cases — such as sickle-cell anemia — will still be made. In the case of cystic fibrosis, missense opens the gates of the cell to a cascade of damaging phenotypes: in 70 percent of persons with cystic fibrosis, the mutation is a deletion of three bases, eliminating a single tryptophan from a critical domain.

Nonsense mutations — which generate silent, recessive phenotypes — jam up the translation machine with unwanted stop codons. To begin translation, the universal assembly line needs to come upon the codon AUG in the messenger RNA. This codon encodes methionine and also the beginning of translation, assuring that all proteins will have methionine as their first amino acid. As Sidney Brenner and Francis Crick — again — first pointed out, the first AUG in a messenger RNA determines how the subsequent sequence will be divided into codons, setting the reading frame of all subsequent codons. If one or two bases are added or subtracted from a messenger RNA, then the translation assembly line will simply roll over the point of deletion or insertion, and continue to put an amino acid in for every three bases. As a result, all the codons from the point of mutation will be out of phase. A deletion of the sixth base in the message that begins AUG GAG CAG CUG AAC, for example, changes the message entirely, to AUG GAC AGC UGA AC. This mutation encodes an abortive, tiny protein that ends when the messenger comes to UGA after only three amino acids. Stop codons that are not in phase with a messenger's first AUG — like the ". . . UG A . . ." of the normal message in this example — have no effect on translation and can therefore accumulate along a gene's sequence, hidden until a nonsense mutation reveals them to the machinery of translation.

Mutations also reveal that the meanings of a gene are not limited to its encoded proteins, complex as those meanings can be. Mutations in noncoding DNA — introns, regulatory regions, and the DNA between genes — may create abnormal forms or amounts of proteins, whose aberrant functions we see as diseases. Thalassemia, a hemoglobin disorder common to people of Mediterranean descent, results from an error in editing the introns of a hemoglobin transcript. In the earlier example involving the tropomyosin gene, a deletion in the regulatory region might produce the truncated, contradictory "No" in place of "Now, here"; a nonsense mutation in the coding region might abort the message at "Ma" so that no meat at all would be made; while a missense mutation might produce the weird "Make some dark mead." A mutation that damaged a splice junction might create the confusing request "Make some light dark meat" or the possibly lethal "Make some dark."

Knocking out the "here" domain of a gene's regulatory region usually just shuts off one gene, but if the regulatory region serves more than one gene in tandem, more than one protein can be eliminated by a single regulatory mutation. For example, our perception of color begins with a network of three sets of cells in the back of the eye. Each set is filled with a different pigment, sensitive either to red, green, or blue light. The genes encoding the red and green pigment genes are on the X chromosome; nonsense mutations in either the red or the green pigment gene generate one of the common color blindnesses, in which a person — usually a man — cannot fully distinguish among colors that differ by the presence or absence of red or green. Such men can still distinguish between blue and red or blue and green, so they see a reduced but still colorful world. In a rare and far more debilitating form of color blindness, a mutation in the regulatory region that simultaneously controls the synthesis of messenger RNA for both the red- and green-sensitive pigments leaves a person facing a blue-gray world, unable to see red, green, or any other color.

A mutation in one of the regulatory proteins that work on the "here" domains of other genes will disrupt the normal discussion among the sets of proteins that determine when and where a gene should be transcribed. This can generate a spectacular case of inadvertent polysemy: the same regulatory protein can have two effects in two parts of the body. For example, if a member of one family of genes encoding DNA-binding, regulatory proteins is mutated or absent, other genes that should be turned off or on are not, and as a result, entire blocks of tissue in the body may be assembled in the wrong place as an embryo develops.

Even completely silent DNA can be the source of new phenotypes. Scattered throughout the genome are short, silent sequences, islands of Cs and Gs that act as spacers between genes. Sequences of -CCG- repeated between thirty and fifty times, for example, are characteristically found between many genes. DNA polymerase must every once in a long while stutter a bit over these runs of CCG, because the lengths of CCG repeats are different from one person to another. These different lengths are inherited from parents as a silent allelic difference that is normally of no consequence. But should DNA polymerase stutter a bit too long, making a sequence of fifty or more CCGs in a row in one place on the X chromosome of an egg or sperm cell, that sequence becomes dangerous. A woman inheriting such a long CCG repeat may be healthy, but as her egg cells are made, the repeat will double and double again, becoming thousands of base pairs long. Her X chromosome cannot carry this pathologically long stretch of repeated DNA without difficulty; its last chromosome band dangles from the string of useless DNA, and genes nearby cannot be properly transcribed. If her child inherits this dangling chromosome, it will be born with the Fragile X syndrome, a common cause of mental retardation.*

These sorts of mutations tell us that our normal health depends on proper conversations among regulatory proteins and noncoding DNA sequences as much as or more than it does on the proteins — like collagen — that get exported by the cell. Noncoding sentences in the DNA language may never get read aloud as protein, but they are nonetheless essential for the maintenance of proper discourse in and among the cells of our bodies.

Translation — both the fact of it and the way it is carried out — makes the complete meaning of a DNA sequence impossible to predict and difficult to fully fathom. The redundancy of the genetic code in the protein-to-DNA direction means we can never be in a position to predict the exact DNA sequence of a gene from its protein. In particular, we may not draw the conclusion that two people with identical protein phenotypes carry identical DNA sequences; the redundancy of the code assures that there will be a vast number of alleles encoding the same normal protein. The exactness and nonredundancy of the genetic code in the DNA-to-protein direction may permit us to predict the amino acid sequence encoded in a transliterated stretch of DNA base pairs, but RNA splicing and editing make this, too, a less than certain thing.

The multiple alleles encoding specific proteins, the sculptural individuality of each protein's folding, the redundancies of codon degeneracy, and the ambiguities of splicing together

---

* The discovery of a runaway simple repeat in the fragile X chromosome opened a successful search for such sequences in the chromosomes of families with other inherited diseases. For instance, the gene that is mutated in Huntington's disease carries a pathological repeat, as does the gene for Kennedy's disease, a complex set of symptoms including progressive muscle weakness and, in men, feminized breasts and reduced fertility.

give the human genome its richness and depth as a text. They also assure that the transliteration of DNA sequences only begins the hugely complex job of reading a genome; indeed, looking at any new sequence of base pairs, no matter how long, is like discovering a written text inscribed in letters that form words whose meanings we cannot fully grasp. Even when the entire sequence of a human genome is transliterated into a string of English letters, that long list of As, Gs, Cs, and Ts will be a lexicon of genomic words and sentences, the meanings of which we cannot predict. The panoply of meanings in any DNA text — from a gene to the whole genome — will be clear to us only when we know, not just its complete sequence of base pairs, but also the full three-dimensional meanings of all its proteins.

4

THE MOLECULAR
WORD PROCESSOR

FOR BILLIONS OF YEARS, cells have transliterated a chosen set of genes into messenger RNAs, translated messages into strings of amino acids, and allowed each amino acid chain to manifest a gene's meanings by folding itself into an interactive protein. The meanings of a gene in a cell's DNA are for the most part hidden from us today, because no linear sequence, whether of DNA, RNA, or amino acids, can reveal the meaning of a protein any more than we might grasp from its letters the meaning of a new word in a foreign language. Proteins, like the meanings of words, can take on additional layers of meaning quite distinct from their primary intention: mutations that permit the transcription of a gene in the "wrong" cell or at the wrong time, followed by others that slightly change a protein's three-dimensional structure, can capture a protein for a completely new use. Even when different proteins are similar in shape, each will still take a unique set of specific actions on a unique set of target molecules. The only way to build a vocabulary of protein meanings has been to discover the three-dimensional structure and the final functions of individual proteins, one by one. This has been a slow, and for the most part unrewarding, enterprise, but we have to learn a new language, and there is unlikely to be a shortcut for the painstaking work of building a vocabulary.

When a stretch of DNA — even a very long DNA, even the

entire human genome — is converted into English letters, it has been transliterated, not translated. Though the translation of the human genome is still at best a distant target, transliteration is at hand. The U.S. government has been giving English letters to the base pairs of a human genome since 1988, when Congress and the president — undaunted by the inability of the National Cancer Act of 1971 to meet its goal of winning the war on cancer by directed research — established the Human Genome Project in a new division of the National Institutes of Health. The project began as one of James Watson's many ideas; to no one's surprise, Watson became its first director. The Human Genome Project is expected to cost several billion dollars — whether in addition to or instead of money that would have been spent for other NIH research remains a sore point for many scientists — but if Watson gets his dream, it will complete the transliteration of a human genome by 2003, in time for the fiftieth anniversary of his discovery of DNA's elegant, functional form.

How will the project avoid assigning false canonical status to one allele of each multi-allelic gene? The human genome would more easily be understood as a legacy of the entire species if the sequences assembled by the project were to come from the DNA of a single anonymous person. The project's goal is less ambitious but just as ponderous as the transliteration of any one person's genome: to place a correct letter — G, C, A, or T — in every position of a consensus three-billion-letter string. A consensus transliteration is not in any real sense a transliteration of *the* human genome. That would have to include the base-pair sequences, imprintings, and methylation patterns of a full, diploid set of twenty-two pairs of maternal and paternal chromosomes plus an X and a Y chromosome. Even then, about one base pair in a thousand is different from person to person, in genes and out. Realistically speaking, there are simply too many available versions of the genome — about ten billion, at least — to sequence them all.

The project has been expensive to set up; working toward a targeted cost of less than a dollar per base pair, it has so far recorded a million or so human base pairs at a cost of tens of millions of dollars. When it is complete, the resulting printout of three billion As, Ts, Gs, and Cs will be a triumph of transliteration and a catalogue of at least one allele for all human genes. The letters of such a catalogue are each about a million times larger than a base pair; this presents immediate problems of storage and access. Like the National Security Agency, the Human Genome Project has turned to powerful computers to store and retrieve its rapidly growing skein of base-pair sequences.

As with any laborious deciphering job, surprises are inevitable: in 1992, three dozen research groups around the world laid out chromosome III of yeast as a string of 315,000 base pairs, less than one tenth the length of a bacterial chromosome and about one ten-thousandth the length of the human genome. When this relatively short genomic excerpt was put through a computer search for stretches that could be translated into protein, all thirty-seven known genes of chromosome III showed up, but so did many other long stretches without stop codons. Some of these are certain to be yeast genes that had never shown themselves by mutation and so were completely unexpected. If larger human chromosomes carry as many surprises, we can expect to find we are carrying, not the current estimate of one hundred thousand genes, but at least four hundred thousand genes, the majority of them unexpected and unknown.

On the way to complete transliteration, the organizers of the project have wisely set some important intermediate tasks. The first is to construct a sequence-based map of the human genome that would locate every known human gene to the left or right of all others in the genome. About a thousand human genes are already known through disease-related phenotypes generated by their mutant alleles. Most of them have been localized to one chromosome and quite a few to one chromosomal band, but our current encyclopedia of genes holds no more than 1 or 2 percent of the genes we all carry; the rest are somewhere in our chromosomes, and we don't know where. To speed the construction of a map that will locate new genes, the project aims to extract short DNA sequences — called sequence-tagged sites — spaced about a hundred thousand base pairs apart, order them in terms of their place in the genome, and link each to the nearest gene. Since each chromosome band has millions of base pairs, scientists seeking the DNA sequence of one of the mapped genes will be able to focus their attention on a stretch of DNA no bigger than a small fraction of a band. At the same time, the tagged sequences will provide an ordered set of landmarks for the localization of genes as yet undiscovered.

Many genes, and certainly all strings of DNA that contain more than one gene, are too large to be purified and studied easily. Instead, scientists store them as a set of DNAs inside bacteria or yeast cells. A collection of DNA fragments that includes all the sequences of a larger DNA is called a library. The term is optimistic; a fragment library is just that, a library of fragments. The DNA-based map will be the concordance that tells us the proper order of fragments in the human genome. The process of lining up contiguous DNA fragments is called walking the genome: each mapped fragment becomes a tool to find the next in line, much the way the aborigines in Bruce Chatwin's *Songlines* remember their way across the great outback of Australia by matching overlapping songs to pieces of terrain. Walking allows very dense, very long maps to be drawn up; the five-million-base-pair *E. coli* genome has been mapped in this way. Walking a library of the human genome allowed scientists to obtain the giant gene for cystic fibrosis and to assemble a map of the human Y chromosome. But even after being fully mapped, large genomes remain tough to transliterate; not even one bacterial genome has yet been transliterated and filed in a central computer bank.

No matter how fine the resolution of genomic maps, they will stay just that — maps of a territory we want to explore, not the territory itself. Furthermore, even the best map of genes will leave most of the human genome in the dark. We can safely bet that no more than 5 percent of the DNA in our chromosomes is transcribed anywhere in our bodies, while the remaining 95 percent is silent. Even if our chromosomes turn out to have four times as many genes as we now think they do, 80 percent of our chromosomes would be silent. Not all of this silent DNA is useless: in addition to the regulatory regions and introns of all our hundreds of thousands of genes, the human genome's silent sequences include the parts of each chromosome given over to proper movement and separation during mitosis and meiosis, as well as about fifty thousand other sequences that DNA polymerase molecules must find to start and stop replication, and the telomeres, a set of repeated short sequences at the ends of each chromosome.*

Despite their occasional capacity to surprise us, the vast stretches of silent sequences in the human genome are boring, and being chosen to transliterate them is no one's idea of a prize assignment. Predictably, most scientists involved in the project hope to work on sequences that speak to the body; especially interesting are those conversing with the brain. A large minority of all the genes we have are active only in our brains. The best way to find them is to treat a preparation of messenger RNAs from the brain with reverse transcriptase to produce a mixture of messenger-size DNAs, each carrying the edited text of one of the tens of thousands of messenger RNAs made only by the brain. Together these complementary DNAs — or cDNAs, as they are usually called — hold all the genomic information a grown brain uses to maintain itself. A few years ago, some members of the project working at the NIH decided to nibble while they sorted through this miscellany of cDNAs: they partially transliterated hundreds of short cDNA sequences made from brain-specific messenger RNAs, leaving the complete transliteration of each gene for some future time. Since each of these short sequences is different and each comes from a brain-specific gene, grazing in this way will eventually tag the full complement of genes expressed in the brain.

Tagging the DNA sequences of brain-specific cDNAs would seem to be a benign goal for the NIH, but it became a source of much controversy. Over Watson's objection, in 1992 the NIH applied for a patent on each of the short sequence tags. Within weeks of the announcement of the government's inten-

---

* The repeated DNA sequences of telomeres each fold into a knot that resists degradation. Telomeres are critical to the long-term integrity of a genome. As we age, our cells have a greater chance of finding themselves unable to divide, because their telomeres get tattered over time. It may even turn out that the loss of telomeres causes our cells to age; if so, drugs that would block telomere degradation might give us a way to slow the arrival of de Gaulle's "shipwreck," the decrepitude of aging bodies.

tions Watson stepped down, refusing to participate in what he saw as a land grab, an illegitimate block to the free exchange of data among scientists in different countries. For Watson — a blunt man of absolute integrity — the project should have been the culminating phase of his life's work, one that began for him forty years earlier with the discovery of the double helix. Instead, because he could not accommodate a political agenda that valued the rights of state ownership above open-ended scientific inquiry, the project lost one of this century's great visionaries.

Two years earlier Watson had written, "The nations of the world must see that the human genome belongs to the world's people, as opposed to its nations." For now, his own nation has failed to get the point. The NIH has stood firm despite initial reservations on the part of the U.S. Patent Office, and scientists engaged in similar genome research in the United Kingdom, Europe, and Japan have pressed it to reconsider. It remains to be seen whether the office will award any patents for incompletely sequenced genes; regardless, it is difficult to understand why even a complete sequence of a normal allele of any human gene should be claimed by any one person, laboratory, government, or nation, for it will always be present and available in the bodies of people all over the world.

In 1993 the Human Genome Project was taken over by Francis Collins, one of the discoverers of the CFTR gene that is mutated in patients with cystic fibrosis. As transliteration and mapping get cheaper and easier each year, it seems clear that the project will continue even without Watson's special blend of political naiveté and scientific vision, although — perhaps inevitably — both its budget and its hopes have been lowered a bit. While it is unlikely that all three billion base pairs of human genomic sequence will be in any one computer by 2003, by then scientists may well have mapped not only the human genome but also the genomes of many of their favorite experimental organisms: maps of the chromosomes of yeast, fruit fly, roundworm, and mouse are filling in rapidly, and maps of the genes of some plants are also under construction. Maps have always been freely available since their information cannot be patented. But because any gene on a map may yield a profitable product, we must hope that in using maps to obtain the expressed genes, scientists involved in the project will refuse to trade their own freedom of expression for a miscellany of patents.

~

Transliterations of human genes will be followed by attempts to understand what they say and do. The tools and techniques used to carry out this work amount collectively to a new machine, what I would call a molecular word processor. The machine — which is really not an artifact but an agglomeration of several technologies devised by microbiologists — is still rather crude; it will be a long time before anyone can transliterate and manipulate the text of the human genome with the ease and convenience of a word processor. But it is a powerful instrument: to use the molecular word processor is to write in the language of DNA, freed from the constraints that natural selection has placed on the possible meanings a cell can put into or draw from its genomic texts. With it, we can insert or read new meanings into existing genes, and construct new contexts in which gene sentences can take on new meanings. We can also create new meanings by rewriting old proteins. To enable us to do this, the machine recombines DNAs that encode the wordlike domains of a gene, writing a new sentence in DNA that yields a novel chain of amino acids. As our tiny vocabulary of protein meanings grows, we will inevitably be drawn to the idea of a molecular literature entirely of our own creation.

Borrowing current computer technology, the molecular word processor now available can be imagined as follows. It has a keyboard of only five letters: the four bases of DNA and the U that RNA takes in lieu of a T. With the letter keys we can synthesize a stretch of DNA or RNA that is dozens or even hundreds of bases long and carries any sequence of our choice. The apparatus that does this imitates DNA polymerase by hooking up a chain of bases, but it does so in the order we choose rather than by complementing the bases on an existing single strand. The result is a wholly synthetic single strand of DNA. To make an RNA, the machine is fed the four RNA subunits along with a desired order of bases. Just as "natural" vitamins are no different from synthetic ones, synthetic RNA or DNA strands have no difficulty being recognized by enzymes. For instance, DNA polymerase will easily convert a single synthetic strand of any DNA sequence into a regular, double-stranded DNA molecule.

The keyboard also has six function keys that carry out the commands to Cut, Paste, Search, Undo, Print, and Copy DNA sequences. Applying the function keys to the human genome and to DNA sequences created with the proper four letter keys, we can find stretches of DNA, remove them from the genome of a cell, splice them to sequences of either nature's making or our own, and transliterate the new genes into English letters as we go. For example, Cut and Paste applied to DNAs from a duck and an orange would give us a novel sequence containing genes from both, one that might be able to make an orange's protein in a duck's cell or vice versa.

The molecular word processor's software is wet: two of its function keys — Cut and Copy — come from microbial enzymes, proteins made by bacteria much like the ones that live in our own bodies, and Undo depends on an enzyme isolated from the kind of virus that causes AIDS. These tools are available to us because bacteria and other invisible life forms use the same genetic code and the same machinery of replication, transcription, and translation that our cells do.

Bacteria are extraordinarily prolific assemblers of genes and proteins: they can divide a few times in an hour. In each short period between divisions, the bacterium has to double everything within itself, including its chromosome. Genes are transcribed and proteins translated at a furious rate; even while the swollen, doubled bacterium pinches itself in two, it is busy making new proteins for the next division. Should life get hard — if the fluid it is in loses necessary nutrients or the temperature departs from a narrow range of comfort — a bacterium will shut down the transcription and translation of most genes, turn on a small set of other genes, and settle down inside a thick wall as a spore in hibernation.

A typical bacterium is a dirigible supported by a framework of proteins and sugars, with a rubbery sac just beneath the shell to regulate the flow of food and waste molecules. Bacteria have little experience of the world except through what is soluble and small enough to reach them in one or another watery environment. Food to them is a set of small molecules, sugars, salts, some vitamins perhaps, and oxygen in some cases to help burn the food. Stripped down to a single circular chromosome a few million base pairs around and an economy that tightly links self-expression to available resources, a bacterium might seem to lead a life of dull perfection. But bacteria are at risk from the same sort of biological extremism that we are: infection by viruses and parasites.

A bacterial virus carries an even smaller piece of DNA or RNA for a genome. As small as a few thousand bases, it encodes little more than the proteins necessary for its own outer coat and the enzymes necessary for its assembly. Inside its coat, its genome is dormant, but if a virus can place its genes inside a bacterium, the perfect if boring life of that bacterium is nearly over. Like a piece of luggage containing a bomb, invading viral genes are innocently handled by the unsuspecting bacterium's DNA-replicating and DNA-transcribing proteins. The result is the rapid proliferation of the virus's genes, messenger RNAs, and proteins, followed by a silent explosion as the bacterium, converted to a bag of viruses, bursts to release them.

The molecular word processor's Cut key is a plowshare beaten from the swords of an invisible war between bacteria and the viruses that infest them. The Cut key depends on a set of bacterial proteins called restriction enzymes, weapons that disarm viral DNA by chopping the genome of an invading virus to bits before it can damage the bacterial cell. Most restriction enzymes find and cut inside palindromes, DNA sequences that read the same on each strand. Acting as DNA scissors, these enzymes are actually pairs of identical proteins, each of which binds to one strand of the palindrome and cuts it.

For example, the common intestinal bacterium *E. coli* is the source of the restriction enzyme EcoRI (pronounced *ee*-koh-are-*one*). When EcoRI cuts inside the palindromic sequence

$$... GAATTC ...>$$
$$<... CTTAAG ...$$

a pair of EcoRI molecules first binds to the same GAATTC> sequence, then each breaks its strand between the G and the first A. This generates a staggered cut in the double-stranded DNA, with each strand carrying the sequence AATT hanging from a double-stranded stretch that ends with a C:G base pair.*

To protect its own DNA from being cut — all short sequences occur frequently in all genomes — every bacterial species that carries a restriction enzyme also carries another enzyme that camouflages the target sites on its own DNA, so that the hand of a restriction enzyme passes over the bacterial genome, remaining free to cut down invading, uncoated DNA. Bacteria that have EcoRI, for instance, cover every occurrence of GAATTC> on their own chromosomes. Restriction enzymes have been purified from hundreds of kinds of bacteria; each cuts DNA at a specific run of four to eight base pairs. Because every restriction enzyme will cut decisively at every occurrence of its target sequence, each enzyme will split a genome into small, precisely edged pieces.

Any two pieces of DNA that have been cut with a particular restriction enzyme have the same trailing, single-stranded palindromic ends, no matter which species they came from. The Paste key uses base pairing to bring together the fragments of DNA created by the same restriction enzyme: the single-stranded loose ends will spontaneously form base pairs, linking the two fragments. Two different fragments of DNA cut by EcoRI, for instance, will connect to each other as the single-stranded AATT> tail of one fragment forms base pairs with the <TTAA tail of the other fragment. Once held together by base pairing, the two pieces of DNA will look normal to the cell, except for a pair of breaks in the backbones of the two strands, displaced from each other by a few base pairs. These can be closed — pasted — by DNA repair enzymes, to create a clean, single piece of DNA. When a new DNA sequence is pasted together this way from the restriction fragments of different organisms' genomes, it is called a recombinant DNA.

---

* That is, the off-center cut

          ... G|A A T T C ...>
          <... C T T A A|G ...

generates two identical, single-stranded ends:

      ... G>   AATTC ...>
     <... CTTAA   <G ...

Recombinant DNAs become interesting when they are introduced into cells, where their novel combinations of regulatory regions and genes form sentences no cell would ever come across in nature.

Bacteria also harbor parasites, small loops of DNA called plasmids that coexist with bacterial DNA and get copied at each generation by the same enzymes. Plasmids do not kill the bacteria they infect but slow them down, like DNA worms in a bacterial puppy. A plasmid's infection of a bacterium may be dangerous for us, because some plasmids encode proteins that break down our antibiotics. Bacterial plasmids that encode resistance to an antibiotic are natural carriers of recombinant DNA. Once such a plasmid is slipped into a cell, it replicates and confers permanent resistance to the same antibiotic on all that cell's descendants, making its presence easy to track.

Making plasmids into carriers of recombinant DNA is another example of the recycling of microbial swords into molecular plowshares. Ordinarily, plasmids are a molecular burden on a bacterium, but in the topsy-turvy body of a patient treated with antibiotics, the circular genome of a plasmid can be a bacterium's lifesaver. An infected person receiving antibiotic treatment often becomes a culture vessel for the enrichment of bacteria that carry a drug-resistant plasmid: as other bacteria die from the antibiotic, resistant ones may thrive. But, using the Paste key, we can insert a DNA of our choosing into a drug-resistant plasmid, then feed it to a culture of bacteria. Only the bacterial cells that take in the plasmid and express its recombinant genes will be able to survive a dose of the antibiotic, thus turning an infectious bacterium's life preserver into a productive parasite of our own making. Bacteria, yeast, and even human cells — grown in dishes, tubes, and vats — have produced a multitude of novel, biologically active proteins from recombinant genes inserted into drug-resistant plasmids, opening a new era of molecular pharmacology.

The Search key also depends on the tendency of two bare but complementary strands of DNA to connect by their base pairs as a double-stranded DNA helix, a property called molecular hybridization. If the two strands of a DNA molecule are separated from each other by heat or by a strong solution of salts, they will each flop around aimlessly. When the temperature is lowered or the salts diminished, the two complementary strands will zip back up again, to form a regular double helix. This coming together can occur only if the two strands match exactly into G:C and A:T base pairs. A synthetic strand of DNA can be used to search for its complement anywhere in the genome: the letter keys of the molecular word processor allow us to make a short stretch of single-stranded DNA that carries a portion of the sequence of a gene we are after. If we chop up the whole genome with a restriction enzyme, separate the pieces into single strands using heat or salt, mix them with the synthetic DNA, and allow the mix to hybridize into double-stranded DNAs, the synthesized probe will hybridize only to its complementary sequence, forming a stable double-stranded DNA of the gene we are looking for.

The function key Undo is the enzyme reverse transcriptase. With it we can get back a DNA sequence from its messenger RNA so that the other keys will be able to work on it. The Undo key can be used to capture in DNA form — for later recombinant editing with the other keys — the set of messenger RNAs produced by any cell from any active gene. For example, we've already seen how the Undo key, followed by Cut and Paste, was used to produce a library of plasmids carrying cDNAs of the thousands of edited messenger RNA sequences produced by brain cells.

To print means to transliterate a DNA's sequence of base pairs into a string of Gs, Cs, As, and Ts. The Print key uses DNA polymerase to produce a nested set of partial-length copies of a restriction fragment. When these copies are separated and displayed in order of size, they reveal the sequence of bases in the original fragment. The trick to printing — as opposed to merely using the DNA polymerase to copy the fragment — is to give it a solution containing all four DNA bases as well as a calibrated dose of a mock version of one of the four, an unnatural caboose that connects to a growing DNA strand and then prevents further synthesis. Whenever the polymerase needs to insert that base — G, say — the new strand has a chance to get a mock G instead and be forced to end at that point. Four separate runs of DNA polymerase with the four mock bases produce four sets of nested partial sequences, each starting at the same place and ending at one or another occurrence of a particular base. When separated by size, the four sets of smaller fragments reveal by their locations the location of each of the four bases in the original fragment, and hence its sequence.

The machinery for printing DNA is quite accurate. But mistakes happen, and in practice the best sequencing laboratories report error rates in excess of one base in a thousand, which is worrisome, given that even simple genes can often run to thousands of base pairs. If even one base of a coding region is misread or skipped, the transliterated sequence will be as different from the real one as if the real one had suffered a frame shift or missense mutation. The most effective, if expensive, way to avoid such errors is to sequence both strands of a DNA fragment. If no errors have been made, Gs sequenced from one strand will always match up with Cs on the other, and As on one strand will match up with Ts on the other. Any deviation from the sequences predicted by the base-pairing rules signals the great likelihood of an error in one of the complementary transliterations, since such an error is far likelier than a mispaired DNA. And these errors matter: any hypothesis about a gene's function built on an erroneous transliteration — even one with just one skipped or added base pair in a thousand — may be wrong.

The sixth function key allows us to make many copies of any string of DNA letters: the Copy key also uses DNA polymerase but combines it with cycles of hybridization to produce huge numbers of copies of one stretch of DNA. When a cell is dividing and all its descendants are dividing in turn, the total number of cells will double and redouble each time a cycle of division is completed. This is a chain reaction: in only twenty doublings from one cell a million cells are born, and in thirty doublings a billion. The polymerase chain reaction, or PCR, forces the two strands of a fragment of DNA into a similar chain reaction of replication. In the presence of primers — short single-stranded DNAs that hybridize to the ends of the fragments — a single restriction fragment molecule will serve to start the chain reaction, which will create billions of copies in a few hours. The root of a single hair or the tiniest drop of blood has many copies of a person's entire genome and much more to tell — with the right PCR primers — than a fingerprint.

The Copy function of PCR depends on the molecular word processor's Print key to reveal the base sequences at the ends of a restriction fragment and its letter keys to chemically synthesize primers of single-stranded DNA that will base pair to these ends. Once primers have been made and tested on a restriction fragment, they can be applied directly to genomic DNA, where they will amplify whatever sequence sits between them. The case the NIH has made for patenting partial sequences of genes expressed in the brain depended on the latent power of PCR; patented sequences could be made into primers and probes that could be used to recover an entire gene's sequence from a preparation of human genomic DNA at any time.

Our molecular word processor is now complete. Let us say we want to make a vat of bacteria into a factory pumping out insulin, a human protein that people need, and one they are willing to pay for. Using Undo, then Cut and Paste on the messenger RNAs of a tissue means recovering a messenger RNA, turning it back into coding DNA, and inserting the coding DNA into a plasmid. Using Undo, we can make a collection of DNAs from the different messenger RNAs extracted from the insulin-producing cells of the pancreas. Since we know the amino acid sequence of insulin, we can use the genetic code as a guide to type in the right DNA sequence to synthesize a set of PCR primers. By applying the primers to our cDNA, we can use the Copy key of PCR to make many copies of the insulin gene's messenger RNA sequence in DNA format. Using Print we can transliterate this messenger RNA sequence, and using Search, we can then find the entire insulin gene in the human genome, complete with its regulatory regions. We can then use Cut and Paste to obtain the complete insulin gene and insert its coding region into a plasmid, next to regulatory sequences that the bacterium will recognize, being careful not to disrupt the plasmid's gene for antibiotic resistance. After recovering a bacterium that has taken in the plasmid by passing a culture through the antibiotic, we can transcribe the insulin gene from the plasmids that must be in surviving bacteria. Finally, we can provide these bacteria with a favorable culture fluid; as they rapidly reproduce, we have a microbial factory — built on a natural version of the Copy key — that pumps out human insulin. Human insulin made this way — Humulin is its trademarked name — has been on sale for some years.

Should we choose to change the amino acid sequence of the insulin our bacterial factory makes, hoping perhaps for a pseudo-insulin with interesting new properties — stability without refrigeration, perhaps — we need only go back to the plasmid, use our keyboard, make a variant DNA, and cycle it through the same bacterial machinery. We could also try for a more elegant solution to diabetes: delivery of the gene for insulin to cells in a patient's pancreas or liver. Keeping in mind the need for complete gene sentences ("Now, here, do this to that"), we could make a coding sequence for insulin follow its own regulatory sequence, or a regulatory sequence that acti-vates transcription in a liver cell, or some third set of regula-tory domains of our choosing. Each variant would be a differ-ent sentence, which may have a new meaning in a cell.

The first electronic word processors had keyboards and screens much like those of today's sleek, portable machines, but they had to be wired to boxes of electronic circuitry that filled a large room. Today's molecular word processor is like the lumbering, vacuum tube computers of the 1960s, but in place of wires and racks of diodes, ours have glass tubes that bring a DNA text to various chemical and enzymatic reac-tions. Cut, Paste, and Undo take place at body temperature in plastic and glass tubes that resemble nothing so much as crack vials. Copy and Search use benchtop instruments that flicker and hum as they work, much like the memory banks of the early computer. The fanciest parts of the molecular word proc-essor are attached to the letter keys, which synthesize short runs of DNA to order, and to the Print key, which reads a DNA sequence from a set of nested fragments; both of these can actually have their own keyboards and video display screens.

Thirty years ago, few computer scientists foresaw cheap portable computers with random-access memories carrying whole books and the programs to write them with; with this history in mind, it is not very daring to predict that the Human Genome Project will bring down the size and cost of the molecular word processor fairly dramatically. Since all the function keys except Search are based on bacterial or viral enzymes, in principle the whole machine could become very small indeed.

~

Beyond its evolving role in the Human Genome Project, today's molecular word processor is used in two other major developments at the boundary of science and public policy: identifying specific DNA sequences in samples of a person's genome, and creating new sequences to insert into the genomes of various kinds of cells. As an identifier, the word processor has been used to tell whether a person is the perpetrator of a crime, the parent or grandparent of a child, or the bearer of an allele of interest to physicians and insurance companies. As a gene recombiner, the word processor has inserted genes into bacteria to produce proteins for sale, into cells for possible injection into a person, and even into an embryo cell whose descendants can become a whole organism.

The molecular word processor can be used as an identifier because everyone's genome is different: about one base pair in a thousand on average is different between two unrelated people. Since even siblings inherit different sets of alleles from each parent, the DNA sequences of any two people — except, of course, for identical twins, who are the offspring of a single fertilized egg — will be unique. Because of its usefulness as an identifier, various models of the molecular word processor are in constant use in district attorneys' offices, personnel offices, hospitals, and insurance companies around the country; the FBI and the Pentagon sport particularly powerful ones.

Very small differences in sequence from person to person can be amplified by the Cut key. Each restriction enzyme cuts only at a specific, short sequence; even though these targets are as long as eight base pairs, a single base-pair difference anywhere in a site may either generate or remove the site from a stretch of DNA. As a result, single base-pair differences between two genomes will sometimes show up as large differences in the size of genomic DNA fragments produced by a restriction enzyme. Differences from person to person in the size of a particular fragment can be picked up by hybridization with a probe to sequences that surround the restriction enzyme's cutting site — in other words, by using the Cut and then the Search keys. Every combination of restriction enzyme and probe will reveal a new pattern of allelic differences — called restriction fragment length polymorphisms, or RFLPs (pronounced *riff*-lips) — between the DNAs of two people. RFLP analysis can be used to distinguish between the DNA of a rape victim and the DNA of sperm left by her attacker; it can also be used to match a tissue sample with its donor and thus identify — or exonerate — a man suspected of the rape.

RFLP analysis is also used to track down the gene responsible for a dominant inherited disease, and to determine — before symptoms appear — whether a child at risk of inheriting the disease received the damaged allele from the affected parent. When a family of many generations is available and members are willing to donate blood samples, their DNAs can be analyzed with a broad spectrum of RFLP probes that seek out allelic differences throughout the genome. If one set of probes picks out differences that are consistently correlated with the presence or absence of the disease through many generations, these probes are likely to base pair to sequences relatively close to the gene that causes the disease. At the request of a family carrying the alleles of an inherited disease, for instance, a doctor can easily and safely sample the amniotic fluid bathing a first-trimester fetus. From the DNA of cells shed by the fetus into the fluid, a lab can tell whether a child will develop Tay-Sachs disease, cystic fibrosis, or Huntington's disease.

In the late 1980s many young Argentineans, who had been kidnapped when their parents were killed during the "great disappearance" of the 1970s, were reunited with their grandparents by RFLP analysis. At first it seemed impossible that the real families of any of these children could be unambiguously identified, but when living grandparents insisted that they knew of their grandchildren's whereabouts, RFLP analysis with a twist provided an unambiguous test. The twist was to examine the small circular DNAs in the child's mitochondria. Because every woman puts hundreds of her mitochondria into each of her egg cells while a sperm donates only its nucleus to the fertilized egg, all of our mitochondria are inherited from our mothers, who in turn inherited them from their mothers. When the mitochondrial RFLP pattern of a child of "disappeared" parents matched the pattern of a grandmother's mitochondrial DNA, a family could be reunited.

To anyone wishing to know a person by his or her genome, RFLP analysis has two drawbacks: it is slow and expensive. Another method of identification using PCR — the Copy key — allows quick and cheap amplification of the short repeats of simple sequences that litter everyone's genome. These silent, stuttering runs have proved to be useful in rapid identification. With some pathological exceptions, like the growth of a CCG repeat in Fragile X families, differences in variable number tandem repeats (VNTRs) are inherited with the same stability as any other allele, so every length of a simple repeat can be treated as a different allele. Amplifying three or four VNTRs by PCR and separating them by size on a gel is a rapid and reliable way to get a record of bands unlikely to appear by chance a second time in anyone else. The result of the test, looking to the uninitiated like the Universal Product Code on a box of cereal, is the hottest ticket in forensic technology: the DNA fingerprint.

DNA fingerprinting is not absolutely unambiguous, but it comes close: only identical twins will have identical DNA fingerprints, and only people who come from very inbred groups are likely to share more than a few VNTR lengths. VNTR analysis is largely the invention of Alec Jeffreys of the University of Leicester in England, a scientist who instantly became an expert on forensics. Jeffreys is probably best known for identifying a disinterred South American skeleton as the remains of the much-sought-after Dr. Josef Mengele of Auschwitz. Jeffreys used VNTR analysis of DNA from the body's bones and from samples donated by his surviving wife and son. Only a fraction of a percent of the DNA from the remains was human; more than 99 percent was from the microorganisms of the grave. Nevertheless, the son's VNTR alleles could all be accounted for in the DNA of the mother and the DNA of the remains, which assured the identification of the body. It is fitting that traces of the genes of this terrible doctor, who killed so many in the name of genetic experimentation and then hid from responsibility for his actions, have been made to speak his name.

Since tracking a disease-related allele requires the cooperation — and the DNA — of many family members, some people have proposed that hospitals take a DNA sample of everyone at birth. But surrendering your genome, whether to a hospital, an insurance company, or an employer, may well mean surrendering your privacy. Once you provide your genome for analysis, your DNA can be tested for the presence of other alleles, now or in the future. Already the FBI and many large police departments have begun to assemble the molecular fingerprints of convicted criminals from DNA samples taken in prison, and the armed forces take and store a DNA sample from every person they recruit. Today, millions of Americans have their DNA on file in a public or private molecular word processor; with the federal government about to take new responsibility for controlling our country's trillion-dollar health care budget, setting the boundary between the voluntary submission and the compulsory surrender of one's genome is likely to become important in the near future.

Perhaps the most dangerous abuses of DNA identification using the molecular word processor will be those built on the correlation of sets of RFLP or VNTR bands in healthy people with differences in complex behavior, especially if the behavior can be shown to be, not only the result of experience, but also the consequence of any number of particular alleles. The human brain, which uses so much of the genome's information, is not likely to play out any of its sophisticated programs — intelligence, memory, humor, creativity — through a single gene, so it would be wise to look skeptically at any behavioral or intelligence tests designed to provide data for correlation to single DNA markers. But it takes only a little imagination to foresee a time when some organization will try to correlate a

RFLP or VNTR pattern with scores on a battery of tests designed to measure intelligence, or sexual preference, and then use this genetic marker to encourage or deny entrance to its ranks.

Assuring that a DNA sample is discarded after an RFLP or VNTR analysis ought to be the legal obligation of every insurance company, hospital, and prosecutor's office. But so far it is not, and because the probes that might be applied to it in later years cannot be predicted, the availability of DNA samples is likely to lead to many surprises, not all of them pleasant. The latent risk of loss of privacy is growing rapidly; keeping and analyzing people's DNA without specific, agreed-upon legal authority is like reading their mail or tapping their phones without a warrant. In France, by comparison, the right to privacy is so guarded by law that scientists and physicians have been prevented from notifying families at risk of inherited disease unless they ask for the information.

Even a single allele in the wrong hands can set unexpected limits on a person's future. If health and life insurance companies are ever permitted to demand a qualifying medical examination that includes family DNA analysis, they will be able to establish whether a prospective client is free of propensities to develop one or another disease, which then allows for the possibility that they would refuse insurance or raise the premium for a person who has no outward signs of disease. It is a bit ominous that the 1992 Persons with Disabilities Act, which provided so much in the way of equal rights and opportunities to people with manifest disabilities, offered no protection against the denial of access to health or life insurance.

Beyond its capacity to restrict access to insurance or health benefits, a test is a powerful instrument of control. Each test's results can be interpreted only by comparing an individual's outcome to the "normal" result, so whoever defines the limits of "normal" can easily stick permanent labels on young children. In *Dangerous Diagnostics*, Dorothy Nelkin offers a taste of the way economic incentives can distort the testing process. When the government introduced programs for children with learning disabilities in the 1970s and schools began to receive funds in proportion to the number of their affected students, the percentage of children called "learning disabled" by school tests increased threefold from 1976 to 1982.

In the absence of clear legal boundaries, we are at risk of developing a de facto national eugenics policy after all, not because we wish to identify and then eliminate people as undesirable members of "lesser races," but because some alleles will be considered undesirable by organizations in a position to limit their replication. We are now on the verge of developing cheap, comprehensive assays for hundreds of allelic differences, some of which will be associated with diseases and all of which will be inherited. The temptation to use this information irresponsibly will be great. It is ever more clear that our laws have lagged behind our technology; they do not recognize the power of the molecular word processor to force individuals and governments into making new kinds of decisions. Dean Hamer, who discovered DNA sequences that are inherited by a subgroup of homosexual men, saw this risk and said so in a blunt coda to his paper:

> We believe it would be fundamentally unethical to use such information to try to assess or alter a person's current or future sexual orientation, either heterosexual or homosexual, or other normal attributes of human behavior. Rather, scientists, educators, policy makers and the public should work together to ensure that such research is used to benefit all members of society.

~

Neither law nor custom stands in the way of using the molecular word processor to make money. Genes have been converted into cash, as pharmaceutical firms sell drugs and chemicals that can be produced only through recombinant DNA technology. Pharmaceutical hormones, for example, are always in short supply; they are difficult to isolate because of their potency in the body. To make a human hormone by recombining its gene into a plasmid costs less than to isolate it from the tissues of animals or recently dead people, and it is less dangerous. Gene-spliced hormones available to date include human growth hormone for dwarfism, erythropoietin for life-threatening anemia, and human insulin for diabetes. Not all have been profitable. Human growth hormone, called humatrope, cost the Lilly company about $100 million to produce; even with federal subsidies provided by the "Orphan Drug Act," it does not appear that the company will make back its investment before its patent runs out.

Several companies have used cells to produce human tissue plasminogen activator, or TPA, an enzyme that dissolves blood vessel clots on contact. TPA and an analogous enzyme from bacteria called streptokinase can both dissolve a new atherosclerotic clot and end a heart attack before great damage is done to the heart itself if they are injected at the earliest moment after a heart attack begins. But human TPA costs ten times more than streptokinase, and the pharmaceutical firms that spent large sums to develop and test recombinant TPA can legitimately recover their investment only if TPA proves to be considerably more successful at saving and prolonging lives than the less expensive bacterial enzyme, without showing any unexpected long-term side effects. Long-term survival rates will only be known after many more years, but recent large-scale, short-term studies in Europe revealed small improvements in survival when optimal doses of TPA were used instead of optimal doses of streptokinase. Since the annual market for the drug of choice to dissolve clots after a heart attack is likely to be in the billion-dollar range, and since so many millions of Americans already cannot cover the costs of their medical care, ongoing cost-benefit analyses of the optimal medical response to a heart attack are unlikely to generate a clear choice between TPA and streptokinase for many years.

Sometimes a recombinant DNA product can be useful to people and profitable to its manufacturers even if it is neither from a human gene nor intended for use as a drug; for example, recombinant genes that improve one or another property of a food crop have been put into many different plants. Though initial experiments of this sort have not caused big problems, they have given rise to periodic sentiment for molecular censorship to keep "unnatural" genes and gene products out of our food supply. Any actual problems to date have stemmed, not so much from the creation of "unnatural" foods, but from the ways in which this new technology has been made to serve the established order of agribusiness. For example, food plants — and tobacco — have been genetically modified to carry genes conferring resistance to various pesticides. These plants certainly do better than "natural" ones in fields heavily burdened with pesticide, but such a use of the molecular word processor does not speed the day when pesticide use can be reduced without loss of profitability.

In another case, companies producing animal feed and antibiotics for veterinary use have pioneered in the development of a gene-spliced, bacterially grown version of a hormone from cows called bovine somatotrophin, BST for short. Periodic injections of BST make cows eat a great deal and give much more milk. Studies done in the mid-1980s at the direction of the U.S. Food and Drug Administration found that milk from cows treated with BST was safe to drink. But the FDA recently acknowledged that BST causes some unexpected problems: the animals eat an extraordinary amount, their immune systems are stressed, udder infections are more common, and, as a result, their milk and milk products are more likely to be adulterated with residual antibiotics.

The FDA has been responding to these new issues raised by BST in a curiously halfhearted way. For more than a decade, four drug companies and their university subcontractors carried out tests on cows to accumulate information about BST and milk from BST-treated cows for the FDA. In 1986 the FDA allowed these companies and their contractors to recoup the costs of their tests by selling the milk and cheese from BST-treated cows without labeling the material as being from animals treated with an as yet unlicensed drug. It would be useful for all of us — and certainly BST's manufacturers — to know which children drank the BST cows' milk and ate their cheese, but FDA secrecy prevents this. If all goes well, we'll never notice, but if a statistical excess of some odd disease ever hits young adults in places like Wisconsin and Vermont, no one should be totally astonished to discover a connection with the milk sold during proprietarily secret early trials of BST.

The social consequences of BST are more immediate and more certain. Farmers in our country already produce so much milk that they can sell it only at a loss; government subsidies keep many dairy farms from bankruptcy. Estimates of the cost to a dairy farmer for BST range from ten cents to a dollar a shot, and the drug has to be administered every few days. This high cost threatens to put some small dairy farmers out of business, meaning that BST is less likely to raise total milk production than to maintain current production with a drop in the number of small dairy farms. BST is thus anything but a miracle drug for regions like rural New York and Vermont; for the rest of the country, the net effect may be higher milk prices. BST may turn out to be one of those advances that takes us in the wrong direction, a drug that costs everyone money but benefits only its manufacturers.

Genes modified in the molecular word processor have also been taken from their test tubes and their plasmids and returned to cellular genomes. I hesitate to call these new sequences genes, though they may be equivalent, because our technology and not nature has written them out in DNA for our own purposes. Although the technology for inserting sequences into a cell's genome is well understood, the eventual consequences of this genetic editing are not. Since we do not know in advance the full meaning of most sequences, we cannot fully predict how the new text will be understood by the cell that has been compelled to take it in.

Inserting recombinant DNA into cells may seem exotic, but it is in fact not very difficult. In 1944 Oswald Avery and his associates at the Rockefeller Institute (now the Rockefeller University) in New York showed that when bacteria took up very pure DNA, the bacterial phenotype changed in a predictable, permanent way. These experiments were the first to demonstrate that DNA — until then a molecule assumed to be as uninteresting as starch, hence given a name based on the deoxyribose sugars of its backbones — carried genetic information. We now know that bacteria and other cells — even human cells — require little convincing to incorporate foreign genes: in various salt solutions they take in all DNA with alacrity, not distinguishing "natural" genes from synthetic stretches or recombinant plasmids.

The name for the insertion of DNA into a cell — transformation — is Avery's, and with slight variations, Avery's chemical solutions perform on cells as well as they do on bacteria. Other techniques also work: if certain crippled viruses are loaded with a new gene, like a Trojan horse they will get it into a cell by infection. One tedious but absolutely certain way to get DNA into a single cell's nucleus — in particular, the nucleus of a fertilized egg cell — is simply to inject it through a very fine glass needle. The nucleus swells as if a mosquito had bitten it, and within hours the cell is expressing the injected gene. More techniques for getting DNA into cells appear every few months; among the more curious is an electric gun that shoots tiny glass beads covered with DNA into a plant cell and the successful injection of purified DNA by hypodermic syringe directly into the muscle of a mouse.

However the DNA gets into a cell, its promiscuous incorporation into the cell's genome assures that from then on a new DNA — whatever its provenance — will be replicated and passed on to all that cell's descendants. Human alleles returned to human cells can themselves be the source of medical treatment, without any bacterial interlocutors. Blood reaches every cell in the body, so it is not surprising that scientists have concentrated on getting the normal, functional alleles of a mutant gene into two kinds of cells from the blood system, white blood cells and the cells that line the insides of blood vessels. In 1989 researchers at the NIH carried out the first approved experiment in humans along these lines. They put a foreign gene that encoded a harmless, easily detected protein into human lymphocytes, then injected the DNA-transformed cells into patients already terminally ill with cancer. The experiment worked, since cells with the marker protein could be found in the patient's bloodstream for many days after the injection.

In 1990 the NIH authorized the same researchers to put new genes in the cells of a young girl sick with an inherited, incurable immunodeficiency syndrome. Due to a mutation, this child's blood cells could not produce the enzyme adenosine deaminase; as a result, the child had little to no ability to make and use antibodies. Her white blood cells were placed in a test tube and infected with a virus whose genome had been recombined with DNA coding for the missing enzyme and its adjacent regulatory sequences. The cultured cells and their descendants made the adenosine deaminase; they were then injected back into the child. She regained a partial immune response quickly, and according to a recent report, after two years "she is able to participate fully in school and social activities." A second child received the same therapy in early 1991 and is also doing well; in 1993 the NIH authorized immediate gene therapy for a small number of newborn victims of the disease. With the discovery of the gene for CFTR, scientists have begun to test DNA sequences that may provide a similar gene therapy for victims of cystic fibrosis. Because human adenovirus naturally infects the bronchial linings, its genome — recombined with a normal CFTR gene — is being tested in cultured cells from the lungs and bronchial tubes. Early reports suggest that this viral carrier can restore CFTR function to such cells, opening the way for clinical tests of gene therapy for cystic fibrosis.

Despite the promise of unique cures for intractable diseases through recombinant DNA therapy, we have to be careful not to step too quickly over the many boundaries we have set on human experimentation. For example, PCR has given us the capacity to diagnose whether an eight-cell embryo has at least one functional allele for the CFTR gene or whether the embryo would grow up to be a child with cystic fibrosis. Currently, embryos that lack a functional CFTR gene are simply not implanted *in utero*. While it may seem a reasonable extension of these studies to try instead to transform such an embryo cell with CFTR DNA and then to implant it with the rest of the embryo, such an attempt would amount to the sort of experiment on a person that most scientists and physicians now choose not to perform.

~

In mathematics, one plus one equals two. In science and technology, one plus one can sometimes come to seven, or any other number, as synergistic interactions emerge from the combination of new tools and discoveries. We entered a magically fertile period of this sort as soon as the science of the Human Genome Project met the technology of the molecular word processor. The results have been considerably less scary than our worst fears but not at all as glorious as the promises made by those who foresaw an immediate revolution in medicine, agriculture, and the like. The science of recombinant DNA has been remarkably conservative in its infancy. It has made only the smallest emendations and changes so far — when measured against the vast unknown stretches of the genomes that fill the living world — but the pace is accelerating. Wheat that grows in herbicide-laced soil has been followed by more than forty species of DNA-modified food and fiber crops; by late 1992 there were more than six hundred tests of such crops under way in twenty countries.

We can expect new meanings to emerge from a new DNA text when commercially viable products have untoward and unexpected consequences in the marketplace. For example, one of the earliest products of genetically modified bacteria to reach the market was the amino acid tryptophan. Overproduced by genes inserted through recombinant DNA, the amino acid was sold as a treatment for insomnia. One batch turned out to contain an unexpected compound as well, two tryptophans hooked together. This new chemical caused serious brain damage to many people before tryptophan was taken off the shelf and destroyed; it seems that the recombinant bacterium made so much of the amino acid that the poison was produced as a completely novel by-product.

The story of bovine somatotrophin also illustrates the problem of unintended consequences. Because cows on BST get udder infections, they have to be treated with antibiotics. Unless farmers are very careful to withhold milk from the market for a day or so after antibiotic treatment, the drug will get into the milk, and children drinking large amounts of milk — or eating a lot of ice cream — will take in antibiotics. That, in turn, would make the consumers inadvertent breeders of antibiotic-resistant bacteria in their intestines; if they were to contract a bacterial infection, common antibiotics might not be effective. This scenario is all too reminiscent of the problem of hospital infections caused by antibiotic-resistant bacteria, and for the same reason: antibiotic-resistant plasmids won't go away just because we have figured out how to use them for our own purposes.

Transliterated human DNA has given us a set of powerful insights into our bodies even without giving up many of its meanings. Recombining genes and inserting them into bacteria will always depend on a very small number of profitable proteins, compared to the hundreds of thousands of different proteins that combine and interact to create each of us; the path to profitable products is never going to lead us to understand the meanings of the human genome. Identifying people by the sequences in their genomes will no doubt grow as an industry too, delicately balanced — if we are careful to maintain the balance — between individual privacy and communal needs. But the sequences used to identify a person, or even to identify a gene in a person, need not be understood at all in terms of their actions in the body. To read the human genome, though, we have no choice: we will have to make our DNA talk to us in words we can understand, words conveying the meanings our bodies have given to each of our genes. That conversation is taking place inside us every minute, but we have only just begun to understand it.

5

TEXTS, CONTEXTS, AND THE

TRANSGENIC STAGE

NATURE HAS BEEN a reclusive author, filling cells with magical effects and hidden intentions. A word processor makes it easy to put sentences together, but a machine cannot give meaning to a set of new words; only the author and the reader together can agree to do that. We will not learn the meaning of a gene by playing with it in our laboratories or computers; we have to put it back in a cell and allow it to show us what it means. This work has begun, as the protein-coding regions of genes assembled by the molecular word processor — in combination with various regulatory regions — have been let loose in one or another kind of cell. The first such gene transformations showed that context is as critical to the meaning of a gene as it is to a word: a gene may mean two completely different things in two different cells or even in the same cell at two different times. To translate the full, four-dimensional meaning of even one gene, we will need to see the pattern of cells in which it is expressed in various parts of the body at various times as well as the consequent interactions of its proteins with other molecules and cells.

When this motion picture of gene expression is played out in full color with a cast of every gene, we call it the life of an organism. Unless we somehow contrive to see that movie in its entirety, we will never accumulate a vocabulary sufficient for a complete reading of the human genome. So far, we have only prepared short, sketchy film clips, focusing on one gene at a time, running them at our own speed over and over until we get each part and subplot down pat, watching single genes play out their meanings in single cells or in whole animals. These short subjects — the lives of a single gene — have already taught us how a gene may kill one cell and force another to live when it should not; how we are built up, front to back and side to side, by a remarkably old set of genes; and how a brain may go bad from the overproduction or absence of a single gene's protein. When we transform cells or organisms with the genes we have first modified in our word processor, we may say that we are transliterating DNA into English letters, writing in the genomic language and, with the help of cells, learning to translate our own scripts a few words at a time.

I spent many years trying to understand a simple DNA sequence, the T-antigen gene of a virus called SV40, a small but subtle infectious agent first identified in the 1950s as a contaminant of cultured monkey kidney cells. Like any other parasite, a virus cannot live except by capturing and subverting a more complex form of life: viruses can make copies of themselves only after first getting into a cell and taking it over. The viruses that plague us take many paths into our bodies. The influenza virus likes the cells that line the nose, the throat, and the pipes that lead from neck to lungs. The polio virus likes the cells that line the intestine, but it will sometimes jump from there to the nerve cells that signal movement from the spinal cord to the limbs. HIV, the virus that causes AIDS, is today the most studied and most feared of human viruses and is among the most finicky. It grows in — and kills — only one of the hundreds of different kinds of cells in the blood. The body's immune system topples from this loss, which is followed quickly by the devastating and eventually lethal consequences of unchecked microbial opportunism.

Some viruses, like HIV, are little shapeless drops of fatty membrane wrapped around a protein coat containing a lethal genome; others, like SV40, are soccer balls of DNA and protein. Viruses like HIV milk their cell slowly, letting it shed millions of viral particles from its surface into the bloodstream. SV40 prefers the blitzkrieg, overthrowing a cell so quickly that in a few days it becomes a bag of viruses. These are not pretty life cycles, but from the viral point of view they are satisfactory examples of form following function, when function is no more than replication. Nothing about any virus is elegant or complex beyond the biochemistry of self-absorption and self-magnification; viruses are the purest product of evolutionary cynicism. Vaccines may lull us into temporary complacency, but we should respect viruses and expect them to be with us for the indefinite future. They, no less than we, have survived to this day; we have no reason to think we will last any longer than they will.

Viruses are the most efficient parasites: they occupy an entirely molecular niche and are free to completely ignore any other part of the body but the molecules they need in the cell they happen to invade. This molecular focus gives viruses their stunning specificity: the meaning of a viral protein can usually be understood only by a limited number of different cells. No amount of SV40 virus, for example, will make a bacterium or any cell of a bird take notice. A few human cells will grudgingly give in to SV40; most will ignore it. In fact, only certain cells of certain monkeys — the kidney cells of African Green monkeys are best — respond with dispatch to SV40's genes. *

To make visible a cell's response to SV40 T-antigen, let us scale up virus and cell, once again making atoms about as big as marbles and turning DNA back into a thick double vine. SV40 wraps and protects its DNA in a knobby spherical shell; enlarged one hundred million times, it would be a decorated balloon about eight feet in diameter, as big across as the thickness of the cell's outer membrane. The rigid viral shell, a mosaic of many copies of a viral protein jigsawed together by the bumps and hollows of their folded shapes, bears an uncanny resemblance to one of Buckminster Fuller's smaller ra-

---

* When two species share a recent common ancestor they are also likely to share susceptibility to more than one virus. It is not surprising, therefore, that African Green monkeys are also among the few higher primates — we are another — to be infected with an immunodeficiency virus; HIV and simian immunodeficiency virus (SIV) are very much alike.

dar domes. Hanging down into it like chandeliers are a few dozen copies of two other viral proteins. These scaffolds pin the virus's circular genome safely in place inside the shell.

Now imagine this virus coming up against a cell in a monkey's kidney. The cell's membrane, eight feet thick and many miles around, encloses a city of cytoplasm, itself surrounding a walled library within, the nucleus. The proteins of the virus's coat stick to the proteins protruding from the cell's membrane, like weather balloons caught in clumps of trees. The cell's response to the presence of an SV40 virus particle is fatally ambivalent. Neither fully welcoming nor sufficiently hostile, the cell draws in the membrane where the virus is stuck to it, bringing the virus into the cytoplasm wrapped in a coat of inside-out membrane. To dispose of this package of litter, the cell begins to chew it up with protein-destroying enzymes. This is just what the virus wants. As its hard shell is shredded, it releases the SV40 genome into the cytoplasm of the cell.

Now seen by the kidney cell not as litter but as a set of genes that has lost its way, the viral genome is carted through the cytoplasm into the cell's nucleus. There, the monkey's chromosomes are a barn-size accumulation of linked bales of carefully folded DNA. The DNA of each chromosome would be anywhere from five hundred to five thousand miles in length; uncoiled and laid end to end, the DNA in the chromosomes of the magnified monkey cell would have a total length of fifty thousand miles. At about five thousand base pairs, the SV40 genome — a zeppelin's necklace — is eventually picked up by molecules of RNA polymerase transcribing their way through the barns of chromosomal DNA. Helped by other DNA-binding proteins that recognize a proper starting point for transcription, RNA polymerase loosens the two strands of viral DNA, then transcribes down the viral DNA until a signaling sequence in the DNA stops it about halfway around the circle. So, in all innocence, the kidney cell opens a gene that will — as soon as its message is properly spliced to be translated by the cell's own cytoplasm into protein — seal its fate.

The viral transcript contains one splice start site and two splice end sites. In the nucleus the cell's sNRPs — artlessly compounding the folly of RNA polymerase — remove one of the two possible introns, turning the transcript into one or another messenger RNA. Carried to the cytoplasm, one messenger RNA is translated into the SV40-coded protein we are interested in, T-antigen. Soon enough, the kidney cell will read this protein as a molecular terrorist's note that says, "Make my proteins, copy my DNA, and when you're done, you're dead." The other messenger RNA encodes a smaller viral protein, unimaginatively called small-t-antigen. Not much is known about small-t-antigen except that it is small, shares a protein domain with T-antigen, and may assist it in some particularly subtle way.

Newly made T-antigen is a DNA-binding protein. The cell, recognizing this, must truck it back into its nucleus. There T-antigen finds its own small circle of a genome among the bales of chromosomal DNA and binds to it at two small response elements — the very regulatory domains that RNA polymerase used to begin T-antigen's own transcription. They are multiple runs of the five-base-pair phrase GAGGC, short passages inside a larger regulatory region of the viral genome called the Origin. By binding inside the Origin, T-antigen starts the process that will reduce the cell from an absentminded host to a zombie.

T-antigen's mere presence at the Origin stops RNA polymerase from beginning any more T-antigen transcripts and shuts down the synthesis of new T-antigen protein. Then, by forcing the cell's blocked, Origin-bound RNA polymerase to flip over, turn around, and transcribe viral DNA along the other half of the SV40 circle, T-antigen generates a new viral transcript. Copies of this RNA are also capable of a variety of different splices, each of which converts a molecule of transcript into a different messenger RNA. Once in the cytoplasm, these different messengers are translated into three new viral proteins. As they are made, they fold up and link to one another to form the hollow shells of new SV40 viruses, empty mine casings waiting only to be charged up with genomes of SV40 DNA.

Back at the SV40 Origin, T-antigen next tricks another set of the cell's own proteins, including its DNA polymerase, into copying the viral genome. From the Origin around to the far side of the genome's circle, DNA polymerase molecules begin to work in both directions until they meet, making two circles from the original SV40 DNA. As soon as T-antigen tricks it into replicating the first SV40 genome, the cell is doomed. In a natural version of PCR, one DNA genome becomes two, two become four, until — in only a few days — the nucleus is packed with millions of circles of SV40 DNA, all spinning off their own copies of RNA to be translated into yet more coat proteins to be assembled into millions of hollow casings. Finally, the new viral genomes start to wrap themselves into tight knots and slip into the waiting coats. The cell is reeling but still alive; T-antigen has one more meaning, and that will finish it off.

Cells are usually at rest in the monkey's kidney; division is a rare event for them. They impose this rest on themselves by making regulatory proteins — p53 is one we have already met, RB is another — to block the transcription of genes needed for cell division. Animals and people depend on these proteins and others like them to keep cancers from springing up; not surprisingly, many cancers contain mutations in one or both proteins. As it happens, T-antigen opens both RB and p53 locks, avidly binding to both of them and distracting them from their cellular tasks. As soon as p53 and RB are taken from their posts by T-antigen, the kidney cell, though eaten from within by the SV40 virus, nevertheless must begin to divide.

This last selfish grab by T-antigen is remarkable for its specificity: p53 was first discovered as a protein tenaciously stuck to T-antigen. Forcing the dying cell to attempt division pumps up its level of DNA polymerase; more DNA polymerase means more viral DNA replication and more viruses, and that speeds the moment of the cell's death. The burden is overwhelming. The kidney cell, deranged by the ever-increasing demand on its resources, breaks open and dies. The library that is the nucleus has been forced to make and deliver millions of DNA circles into the giant balloons that fill the few square miles of cell surrounding it; library and city both are dying of the unending outpouring of new SV40 viruses even as the cell, deluded, prepares to divide. To the monkey whose kidney may have just lost a few cells, nothing much follows except perhaps the mild discomfort of a chronic kidney infection. But for the new SV40 viruses, released from a dead kidney cell into the full bladder of an infectious but otherwise active monkey, these events are the culmination of a carefully honed plan to reach a jungle full of other — as yet uninfected — monkeys.

In the context of a monkey's kidney cell, the few thousand base pairs of DNA that encode T-antigen have no fewer than five different meanings. First, the sequence itself becomes a double entendre as soon as it is spliced into alternative messenger RNAs. Second, once T-antigen is returned to the nucleus, it finds the regulatory domains of its own gene and, by blocking the "here" sequence in the viral Origin of replication, quickly shuts off the production of its own messenger RNA. T-antigen then obliges the cell to make a new transcript instead, starting again from the Origin and again going halfway around the circle, but in the other direction. This — the third meaning — makes the cell produce millions of empty shells. Fourth, by binding to the viral genome's Origin of DNA replication, T-antigen forces the monkey cell's replication enzymes to make copies of the viral DNA circle, providing new viral genomes for the waiting, empty coats. Fifth, by binding to the cellular proteins that usually block cell division, T-antigen puts the infected kidney cell through a last burst of growth before it dies.

The monkey's infected kidney cell must accept SV40 T-antigen's multiple meanings, even at the cost of its own life. Such an outcome is dramatic, and the death of a monkey cell by SV40 infection may leave one with the sense that the cell was simply mechanically obeying a single set of precise instructions. Such a conclusion would be premature: in a different cellular context, T-antigen's meaning can be equally clear but entirely different. SV40's family name, SV, merely tells us that it is a simian virus; its surname, 40, has a more interesting provenance, which points to the second way a cell can interpret T-antigen. SV40 was the fortieth virus to be found in a wide-ranging search for infectious agents lurking in the cultured cells of monkey kidneys. This was no Nabokovian butterfly hunt; it was a somewhat belated scouring of cell cultures that had been used to grow the polio virus for conversion into vaccines. A small number of those cultured monkey kidney cells would always fill with holes and die, even before they saw any polio virus. These cultures were used anyway — there were no others to turn to — but not before being tested by periodic injection into various laboratory animals: mice, hamsters, and the like.

The search that led to the discovery of SV40 began when an extract of cultured monkey cells injected into a hamster caused a lump to grow under its skin; subsequently, it became clear that the contaminating virus — SV40 — could make tumors in hamsters and mice. Decades later, we can say that the considerable number of people vaccinated for polio with preparations containing SV40 did not come down with tumors any more frequently than those who did not receive any SV40 with their vaccine. This stroke of good fortune has been seen as such by the government agencies that license the sale of pharmaceuticals prepared from the cultured cells of human or primate tissues. All such preparations must now be rigorously tested for the presence of novel viruses well before any new drug or vaccine made from them reaches the market. But all the care in the world cannot guarantee that we will never again discover a latent virus in a pharmaceutical made from cultured primate cells; that is one good reason to welcome the molecular word processor and, with it, the ability to produce vaccines and other recombinant pharmaceuticals in bacterial cells.

Because it could make hamster cells into cancer cells, I took on SV40 as my chosen foe when I enlisted in President Nixon's War on Cancer. I was not a lone soldier; spurred on by the lucky outcome of the vast inadvertent experiment with SV40 that polio vaccination had become, many molecular biologists embraced SV40's ability to cause tumors in mice and hamsters, sure that this very small virus would be — in the jargon typical of the Vietnam era — a good weapon. Its simplicity was its main attraction; such a simple key might well be the best way to unlock the complicated problem of cancer.

Few of us would have predicted then that twenty years later the war would still be fought along pretty much the same lines or that the casualties would be even heavier today. The SV40 key got stuck in the lock back then: when SV40 was first elevated to the status of a model system for cancer, p53, which ultimately provided the explanation for SV40's ability to cause a cell to become a tumor, had not even been discovered. Now that we understand the role of proteins like p53, at least we have a good idea of how the lock works, if not how to slam the door shut again in a cancer cell.

Binding to p53 is the last of the five meanings that T-antigen combines to kill a monkey kidney cell. In a different species of cell — one that can draw from the protein no other meaning except that last one — SV40 T-antigen can mean, not the death of a single cell, but the death of an entire organism. Because monkey and human cells draw all five meanings from SV40 T-antigen, they will both die of the virus, precluding this consequence of a long-term interaction between T-antigen and p53. The species of cells that SV40 virus can transform into cancers are ones in which the virus's T-antigen cannot bring about its own genome's proliferation but can still inactivate the cell's p53: SV40 virus cannot kill, but will transform, the cells of a mouse, a rat, and a hamster with about equal efficiency. Since the other four meanings of T-antigen are played out elsewhere on the viral genome, an isolated gene for SV40 T-antigen will be as effective as the whole virus at transforming cells of these species, and by itself it will also transform monkey cells, even the cells of a monkey kidney. Cells of birds and reptiles have p53 molecules that cannot recognize T-antigen and polymerases that cannot begin to copy SV40 DNA, and so the virus means nothing at all to them.

The capacity of SV40 T-antigen to transform the cells of many different species of mammals suggests that the p53 genes of these species encode versions of p53 that share at least one specific three-dimensional shape or domain — the one that T-antigen can recognize. In fact, the amino acid sequences of a few domains of p53 are almost identical in primates and rodents. This is a clear sign of p53's long history as an important molecule, since it has been at least fifty million years since mice and men — and women — last shared a common ancestor. The conservation of domain structure across so many years explains how T-antigen can form the same tight connection to the p53s of so many species and why in every case the combination subverts p53's ability to keep cells from dividing.

How does SV40 T-antigen make a mouse or hamster cell into the parent of a lethally abnormal growth? Usually, with no explosion of newly made SV40 DNA in the way, a rodent cell exposed to SV40 T-antigen will divide a few times and then, after it digests the irritating pulse of T-antigen and its hapless viral genome, go back to sleep. But every so often the incoming SV40 DNA finds its way into a mouse or hamster nucleus and is recombined into one of the cell's chromosomes. Once this happens, the SV40 genome is no longer vulnerable to the cell's security forces; it is seen by the cell as just another bunch of genes somewhere in its giant genome, genes that will be copied each time the cell's own DNA is replicated.

Unable to make itself fully understood in such a cell, SV40 T-antigen will never have the chance to begin the autonomous duplication of its genome. But its gene can — and will — continue to make T-antigen even after having been incorporated somewhere in the cell's genome, and that T-antigen will force its cell into division in turn. Each time a T-antigen-driven cell divides, the viral gene is copied along with the cell's genome, and as soon as division is complete, both daughter cells are subject to T-antigen and must divide again without delay. The descendants of a cell that has taken T-antigen's gene into its chromosomes can never escape the goad of T-antigen. It stimulates every generation of T-antigen-addled cells to make a new generation of cells in turn. In the body of an adult mouse or hamster, where cell division is very tightly controlled, this is not a healthy situation.*

When a growing nest of unstoppable cells crops up in our body, we call it a tumor. The medical specialty dealing with tumors is called oncology, so a viral gene that converts a normal cell to a tumor cell is called an oncogene, and the

---

* In some tissues, normal cells rarely divide; in others like skin and the lining of the gut, cells divide as a matter of course, but for every cell that is born another dies, shed into the gut or off the body as dandruff. When the obligation to divide is inherited by the descendants of a single cell for many generations, not linked to equivalent cell death, it must produce an ever-larger mass of cells.

process that led to the untoward discovery of SV40 is called oncogenic transformation. Like bacteria in rich broth or primed DNA in a polymerase chain reaction tube, tumor cells grow exponentially. Doubling in about a day, descendants of an SV40-transformed cell will form a visible nodule in a dish in only a few weeks. If a transformed cell is placed in the body of a susceptible mouse or hamster, a tumor appears in about the same time. A million transformed cells — a pinhead clump, too small to feel beneath the skin — will accumulate from one SV40-infected hamster or mouse cell after about twenty doublings, which can take as little as three weeks. Time for another ten doublings — a week and a half — is enough for these million cells each to divide ten times more, producing a billion-celled, pea-size, growing nodule of the sort we are all encouraged to report to our doctors.

T-antigen's inactivation of p53 has taught us that a cancer may begin with the corruption of a single protein of the cell and the subsequent uncalled-for growth of a cluster of cells somewhere in the body. But within that cluster, a selection for the most abnormally vigorous mutant cells goes on constantly: the rest of the brakes on the growth of cells in the body — cell-cell contact, limited energy sources, necessary hormone signals — all select for, and enlarge the fraction of, any rare cell in the initial tumor that may lose sensitivity to these controls by virtue of a second, third, or fourth mutation. In this way, a benign lump or polyp that is constrained from pushing through the layers of tissues that separate the organs of the body will soon harbor within it a more deadly, malignant variant, whose descendants will travel deeper into the body, throwing off tiny microclusters of cells that become secondary tumors throughout the body.

A few hundred growth-controlling genes have been identified as likely targets for mutation in one or another human tumor; cells of malignant colon tumors, for example, are usually mutated in four or more of these genes, and one of them is almost always the p53 gene. Remarkably, a large fraction of tumor-associated human genes — called cellular oncogenes — are closely related to viral oncogenes. It appears that most viruses that cause tumors do so either because in the distant past they captured a mutant form of a cellular oncogene or because — like SV40 — they order a cell to make a protein like T-antigen, which then inactivates one or more of the cell's growth-controlling proteins. Either way, the messages that tumor viruses bring to a cell interrupt the cell's hormonal and genomic dialogue with its neighbors, forcing it to divide and preparing its descendants for the accumulation of further mutations. The oncogenic genes of tumor viruses are intimately related to the growth-controlling proteins of the cell, which explains why cancers can be caused by either viruses or mutagens like cigarette smoke, ultraviolet light, or ionizing radiation: all these agents can change the fate of a cell by bringing on the inherited loss of one or more growth-controlling proteins.*

Once one of our cells has sprung completely loose from the control of proteins like p53 to become a malignant tumor, we have no way today to put its descendants back into a regulated state. Instead, we must remove the initial mass of runaway cancer cells with a knife and try to kill escapees and holdouts with radiation and chemical poisons that damage their DNA. This is a woefully primitive and risky way to deal with the millions of cells in a tumor, and while such treatments may kill most or even all of the tumor's remaining cells, they also may damage the genes for other growth-control proteins in other parts of the body. For the tumors we cannot prevent — the ones that will always occur by random mutation — safe, certain treatment will become available only once we can repair or mimic specific growth-control genes and their proteins in the cancer cells themselves. Until then, we must muddle through as best we can: the best defenses against cancer remain preventive measures such as avoiding mutagens

---

* T-antigen is a member of the family of DNA-binding oncogene proteins. Other oncogenes make counterfeits of cellular proteins that signal the nucleus to begin preparing for cell division. One, for instance, mimics a membrane-bound receptor for a growth-stimulating hormone; a cell that has fake, oncogene-coded receptors on its surface is tricked into acting as if it were bathed in growth-stimulating hormones whether or not any hormones are actually present.

like cigarette smoke and getting checked regularly for small, persistent lumps.

The gene for the T-antigen protein has no single set of meanings; it is an example of molecular polysemy. A mouse cell and a monkey's kidney cell provide the same T-antigen gene with two different contexts, changing T-antigen's meaning in the starkest way. The monkey cell gives the virus a context for self-expression and survival. The moment a monkey cell reads the T-antigen gene of an SV40 genome, it is doomed to convert itself into raw materials for the satisfaction of T-antigen's demands. But if the gene for T-antigen becomes part of the genome of a mouse cell, it is no longer a matter of the cell's being taken over by an unstoppable viral genome but of a cell's being overtaken by its own latent capacity to grow, a body taken over by unstoppable cells. In one context, a cell dies; in the other, it lives and proliferates until its own unnatural vitality threatens to destroy the body of which it had been a part. T-antigen's ability to cause a tumor in a mouse serves its viral genome no purpose; a monkey cell and a mouse cell are like two readers who interpret the meaning of the T-antigen gene in different ways.

Both readings — by the monkey cell ready to die and by the mouse cell ready to be born again as a tumor — are exact, and neither is more nearly correct than the other. Their exactness lies in the three-dimensional shapes of the T-antigen protein's domains. Domains are interdependent contributors to the overall meaning of the molecular sentence that is a protein. Nevertheless, like words in a sentence or the fingers of a hand — recall the handlike DNA polymerase protein — a protein's domains carry their precise meanings separately. Fragments of the T-antigen gene, if addressed properly with the appropriate regulatory sequences, will yield truncated versions of T-antigen carrying some but not all of its domains. When put in a cell, some of these DNA phrases will still be able to carry out one or more of T-antigen's many activities.

As T-antigen comes into intimate contact with different molecules — its own Origin DNA, the monkey or mouse versions of DNA polymerase, p53, RB, and RNA polymerase — each molecule finds different T-antigen domains, and each meets the T-antigen molecule in a special way as the two come together in a complex. What is true for SV40's T-antigen will certainly be true for many human proteins: meanings are added to each protein by the other proteins it joins, and whatever makes the palette of protein interactions different from cell to cell will multiply the contextual meanings of all the proteins in each complex. As the behavior of T-antigen suggests, the ability to understand such layers of meaning will be required of all sophisticated readers of DNA.

～

Although each domain means one clear thing, there is no rule in the DNA language that prevents two or more domains from saying the same thing: two proteins of different amino acid sequences may fold into almost the same three-dimensional domain. The terrorist's note of a virus like SV40, for example, is written in slightly different ways by different viruses, but they still have the same meaning in the right cell: polyoma is a virus that cannot grow in monkey cells but can do very well in the kidney cell of a mouse. Polyoma makes a T-antigen that does to a mouse's kidney cell precisely what SV40's T-antigen does to a monkey's kidney cell: by commandeering its gene expression and DNA replication, each protein enables its incoming virus to kill a host cell in very short order. Both T-antigens bind to an infected cell's p53 protein: even though the genes and their amino acid sequences are not similar, the three-dimensional shapes of these two T-antigens are similar enough to be recognized by the same p53 proteins.

The capacity of genomes to converge on the same meaning through the evolution of different amino acid sequences that fold into almost identical three-dimensional domains enriches the genomic language's ability to express subtle nuances. Just as a trained scholar may be able to recover the provenance and full meaning of an ancient text by comparing many slightly different drafts done at different times by different people, a comparison of similar domains in proteins that operate in different contexts can show us the domain's range of meaning. There is a striking similarity, for instance, in the way two very different viruses, SV40 and polio, make their outer coats. Both viruses encode a coat protein that folds into a structure resembling a partly opened book standing on its bottom edge. In both cases, five of these open books come together at their spines to form a five-sided dome, and in both cases the five-sided domes assemble into twenty-faceted, roughly spherical coats. SV40 and polio share the domain that folds into a set of hooks and snares that can assemble by fives into domes and then by domes into a hollow shell. Yet SV40 and polio share no base sequence, nor any amino acid sequence, in their coat-protein genes.

Why are the shapes of these domains more similar than their amino acid sequences? When two living things bear a likeness, either both are descended from a common ancestor who shared that characteristic or they have developed it independently in response to similar needs over a long time. In the former case, similarities are said to be homologous, in the latter, analogous. Homologous — but not analogous — similarities allow us to reconstruct evolutionary relationships among today's living things. Are domain similarities homologous or analogous? There is no rule to tell us in advance, but either answer is informative. If many domains turn out to be analogous — the way the separately evolved eyes of a protozoan, a scallop, a squid, a fly, and a person are analogous — that would teach us that the vocabulary of functional domains is a small one, perhaps small enough for us to learn it in its entirety. If we find that the preponderance of domain similarities among proteins are homologous — as we already know many to be — then our ever-growing vocabulary will continue to provide us with the historical context for DNA's current meanings.

One DNA-binding domain critical to the early development of animal embryos has been at work for billions of years. This conserved domain is made from a tight helical cylinder of amino acids that trails a floppy string, with a second cylinder lying crosswise on the first; it looks like one firecracker lying on top of another, tied together by their fuses. The helix-turn-helix domain, as it is called, fits nicely into the wide groove of a DNA molecule. There, specific amino acids recognize specific DNA sequences; when the right DNA sequence is found, the helix-turn-helix domain binds its protein tightly to the

DNA, leaving the other domains of the protein available for interaction with other proteins. Helix-turn-helix domains are so ancient, they are found in proteins that regulate the activity of genes in bacteria and their viruses and in proteins that lay out the futures of cells in very early human embryos. In all cases, from bacterial viruses to the embryos we all once were, the same domain is doing the same thing: helping to turn transcription on or off by binding tightly to the regulatory region of a gene. In animals, the helix-turn-helix domain is so good at what it does, and what it does is so necessary to the formation of an embryo, that the sequences encoding it have been preserved inside a multitude of different genes with only a small amount of mutational variation since the last common ancestor of flies and people grew from a fertilized egg and swam in billion-year-old seas.

Flies, people, and many other multicellular animals have mirror-image right and left sides, with fronts and tops different from backs and bottoms. Bilaterally symmetric animals got that way because as early embryos they were built up, head to tail, from bilaterally symmetric segments; every segment — from the head, thorax, and abdomen of a fly to the head, limbs, and ribs of a person — kept its bilateral symmetry as it developed. All bilaterally symmetric animals use a family of helix-turn-helix proteins to begin this body plan by carving crosswise bands from the mass of early embryonic cells and converting a ball of cells into a Michelin man of stacked segments. The family of genes that causes the head-to-tail segmentation of early embryos was discovered many years ago in the fruit fly as a set of recessive mutations that led to the displacement of organs so that, for instance, a leg might grow from the head in place of an eye. Such massive rearrangements of the body plan are called homeotic, from the Greek for "similarity disease." Each homeotic gene encodes a DNA-binding protein of many domains. Some domains interact with other proteins, others restrict the protein's DNA binding to specific short sequences in the regulatory regions of other genes. But all homeotic genes share one highly conserved DNA sequence encoding the ancient DNA-binding helix-turn-helix domain: the homeobox.

The patterns of transcription of homeobox genes are strikingly similar in the very young embryos of a fly, a frog, and a mouse. In each, families of homeobox genes are sequentially transcribed for short periods of time, starting at the head of the early embryo and sweeping down to its tail. Each family of homeobox genes encodes a set of gene regulators: the earlier, headmost homeotic proteins activate the transcription of later, tailward ones; the tailward ones shut off the synthesis of their activators. Such feedback loops generate pulses of gene transcription, like the pulse of transcription of T-antigen in an infected monkey cell. Rather than leading to cell death in this case, pulses produce bands of cells containing different homeobox proteins, striping the early embryo from front to back. Nuclei in each band, filled with a band-specific set of gene-regulatory proteins, begin different cascades of gene regulation, making their cells differentiate until they become a particular segment of the body. Our brains are made of about a half dozen of these early segments, and a glance at our skeleton tells us that the rest of our bodies are made of segments as well.

Homeobox genes have a special relationship with the chromosomes in which they sit. In the fly's genome and in our own, the genes for each family of homeotic proteins are arrayed on a chromosome in the same order as the positions of the bands they direct the embryo to form. The gene for the earliest, headmost homeotic protein is at one end of a chromosome and is transcribed first, the gene for the tailmost homeotic protein is at the other end of this region of the chromosome and is transcribed last, and all the other homeotic protein genes lie in the order that can be predicted from the sequence and position of their expression in the embryo. These beautiful arrays of genes, their chromosomal positions conveying their intentions even before transcription, stand as a major exception to the general rule that the genome is a lexicon of topics linked only through gene regulation.

As new species of multicellular animals emerged from old, the number of homeotic genes in a family, and even the number of families, has increased. For example, in the fly there is only one family of homeobox genes; in simple vertebrates like the lamprey, two; and in mice and humans there are four.

Tandemly repeated, slightly different genes such as those in the homeobox families accumulate from the slowest but most important kind of DNA movement in the genome: the periodic, accidental duplication of a long stretch of DNA. Rarely, two chromosomes fail to exchange DNA symmetrically during recombination, so one germ cell gets two successive alleles of a gene and the other neither. A sperm or egg with no allele is clearly mutant and likely fatally so, but one with two copies is simply overendowed and usually fertile. In an embryo, the extra copies of genes in a stretch of reduplicated DNA will be redundant, as one allele from each parent takes care of the gene's business.

But redundancy turns out to be a great motor of genomic diversity over time: animals with two versions of a gene on each of their chromosomes will survive despite the accumulation of random mutations in one version as long as the other retains its original, functional sequence. Mutated second genes will usually be of no use, but they expose cells of the body at various times to mutant, variant proteins that need not retain the function of the original. Should this second function be advantageous to the survival of an organism, it will tend to be retained through the generations at the expense of organisms in the same species that lack it. Our color vision, for example, depends on a pigment gene on the X chromosome that duplicated only about thirty-five million years ago, giving us genes for pigments sensitive to red and green light in addition to a gene for blue-sensitive pigment. All descendants of the Old World higher primates, including us, have these three pigment genes; New World monkeys still have only the blue pigment and one other. They see the world the way our common ancestors did long ago, in blues and brownish greens, pretty much the way a color-blind person sees it today.

Because homeobox genes direct the organization of the embryo, each successful duplication has led to a new wrinkle in the basic body plan. The homeobox family expressed in head segments of developing mammalian embryos, for example, has extra, unique genes not found in other animals with backbones, suggesting that an old duplication of segments at the front end of an embryo provided the cellular material for the differentiation of a mammalian brain. There is no reason to think the last gene duplication has occurred; based on the past, it is a safe bet that in the distant future new species will differ from current ones in part because of the novel activity of newly duplicated homeotic genes. Nor are homeotic genes the only ones still being slowly multiplied: duplication of the green-pigment genes happens all the time on the X chromosome, and a few percent of people have four color-pigment genes as a result. In the very long run, their descendants may well see a more colorful world than we do.

~

Homeobox genes, and other genes such as those that encode T-antigen and $p53$, challenge the ingenuity of prospective readers of the genome. We can fiddle with the sequences of these genes in our molecular word processor, but what good is that if we cannot see the genes in action? These genes have subtle meanings that cannot be fully displayed in a single cell; they require a stage the size of a whole organism and its entire lifetime to play out their roles. Bacteria are not the answer. Inserted into bacterial plasmids, human genes can be grown and harvested or made into simple protein factories, but a human protein cannot completely express itself among a rag-tag band of bacterial proteins, few of which will notice it. Nor are cell cultures sufficient; a gene of interest may kill or transform a single cell, but it cannot show what it might have done in an entire embryo. To learn how subtle genes work, we have no choice but to get them into every cell of an embryo and watch their effects over the organism's lifetime.

Genes that begin in the lab and are inserted into a living genome at the earliest possible stage in its development were given a name in 1980 by a biologist at Yale, Frank Ruddle: transgenes. Biologists have been putting genes onstage this way for about twenty years. The easiest to make are plants with transgenes; the genome of a plant cell is far more malleable than any of ours. A skin cell in a dish, for instance, will either grow as a skin cell, transform into a skin tumor, or die, no matter what hormones we present it with. But a cell from the stem or root of a plant, in a dish by itself, can be tricked by simple mixtures of plant hormones into thinking that it is a fertilized egg cell. It will start a plant from scratch out of its own copy of the species' genome, dividing and sending out roots and shoots, leaves and flowers, forming itself into a little plant right in the dish. For uniformity and convenience, most reforestation programs now plant evergreen trees cloned in test tubes from single cells; it is actually easier and cheaper to clone some plants than it is to wait for them to fertilize themselves and provide seed to plant.

With as little effort as it takes to separate plant cells from each other and get the DNA in — this is done by first stripping the cell's wall off, then injecting the DNA or slipping it in by transformation — transgenic plants can be started, planted, and harvested. To begin the cloning sequence, a gene from a molecular word processor is injected into the nucleus of a cultured plant cell before the cell sees any hormones. Then hormones are added, the clonal sprout is cultured in a dish, then in the ground, and soon it becomes a complete, transgenic plant with a copy of the injected gene in every one of its cells. Since the pollen and egg cells of a transgenic plant will carry the transgene, DNA has to be injected only once to begin a bed, or a whole garden, of transgenic plants. Thereafter, they can be bred to one another to make a new strain with whatever feature the transgene confers. Transgenic plants accomplish what natural selection takes much longer to do by gene duplication: they provide a laboratory for the testing of extra, variant copies of plant genes.*

Transgenic plants by the hundreds have been made and grown for their resistance to herbicides. These plants — a mixed message from the molecular word processor — improve crop yields, but they also encourage the use of herbicides to

---

* Some years ago, for instance, scientists discovered how to grow a transgenic plant that could create an antibody. Knowing that an antibody is made of two different proteins that fold into a single active complex, they created two transgenic tobacco plants with DNAs made from the messengers for two proteins. Each made one of the proteins in its leaves; as expected, neither protein had any antibody activity. When the two transgenic plants were crossed, offspring that contained both genes made both proteins, and their leaves filled with active antibodies.

kill competing plants in the field, a practice that has obvious drawbacks. Most important, high doses of herbicide result in the eventual overgrowth of herbicide-resistant weeds by the same logic of natural selection that makes penicillin treatment the cause of penicillin-resistant infections and spraying with DDT a way to generate swamps full of DDT-resistant mosquitoes.

Recently, the Food and Drug Administration decided that transgenic plants could be the source of foods and drugs for general consumption without prior government testing and left the decision of whether to test such plants to their growers. Since many common food plants — potatoes, for instance — contain poisons in parts that are routinely discarded, the government is taking an inexplicably optimistic view about the potential for subtle toxicities in transgenic foodstuffs. Nor is the government requiring that manufacturers and growers label transgenic plants so that consumers would know what they were buying. Like it or not, under current government regulations unlabeled milk, ice cream, and cheese from BST-treated cows and fruit, cereal, and vegetables from transgenic plants will soon be in our stores and on our tables.

∼

It doesn't make much sense to put T-antigen, p53, or a human homeotic gene into a plant cell. Plants are built up differently from you and me: they have no front or back, differing only from top to bottom. They cannot move around, nor can any but a few of their wooden-walled cells move at all. After every cell division, most plant cells are fixed in place for their lifetime. Like chimneys growing taller by the addition of bricks, the roots and shoots of a plant grow by the division of bands of cells, which are constantly being displaced upward, downward, and outward by the tubes of woody cells they leave behind. Despite their immobility, plant cells communicate with one another as our cells do, by sending proteins and small molecules called plant hormones through hollow tubes filled with fluid. Under the direction of plant hormones, the dividing bands periodically lay down clusters of cells that later undergo meiosis to become the pollen and eggs of a flower. Plants hang their genitals out in the open, and pollen is the one part of a plant that can move; whether by insect, wind, free fall, or fire, a plant must make sure that its pollen gets to an egg cell to start a new plant. Proteins such as T-antigen cannot perform in any sensible way in a plant; the genome of an animal is their only legitimate stage.

There are two ways to generate strains of transgenic animals. The first uses a cancer cell as a vehicle. Mice — and people — sometimes suffer from a rare form of cancer called embryonal carcinoma. Its name conveys the nature of this tumor, which springs from a deranged cell in the tissues that make sperm and egg. Embryonal carcinoma (EC) cells are cancerous; instead of becoming a sperm or egg cell that will contribute to an embryo, they make a horrible kind of tumor, full of differentiating cells that form mockeries of various tissues interleaved with nests of rapidly dividing, undifferentiated EC cells. The undifferentiated cells grow well in a dish, and like many other cultured cells, they can take in a recombinant plasmid and allow its genes to be incorporated into their genome. Unlike almost all other cultured animal cells, though, an EC cell — whose parent cells were about to undergo meiosis — can recombine one of its own genes with a similar sequence that is brought in on a recombinant plasmid. In one more botched attempt at the meiosis it will never get to, an EC cell may eliminate one of its own genes, replacing it with a laboratory version. This noble failure is called site-specific recombination. Through it, a recombinant gene can go directly from the molecular word processor into the EC cell's genome, either knocking out the EC cell's version of the gene or replacing it, giving a newly made variant the perfect stage for a full performance.

Getting the DNA-transformed EC cells into a mouse embryo means interrupting the normal course of mouse development soon after fertilization. Embryos no older than a few hours, and no bigger than a few dozen cells, are dislodged from a recently mated female mouse. Each embryo is a tiny, hollow ball of cells. A transformed EC cell is inserted through a needle into this hollow ball, and the mouse embryo, now bearing a cell that would overwhelm and kill any adult mouse in a few weeks, is returned to the uterus of another mouse. Something amazing happens next: the EC cell starts to divide, but as it does, it ceases to be a cancer cell. Surrounded by the dividing cells of the embryo, the progeny of the EC cell become normal tissue cells, and the mixed ball of cells grows into a mosaic of embryo cells descended both from the fertilized egg and the EC cell. Such a beast — it puts the mythical chimera to shame — has not two but four parents, two of which had conceived the mouse that had carried the embryonal carcinoma in the first place. Tetraparental mice are fertile unless the EC cell's recombinant gene commands otherwise, and so are their descendants. When a son and daughter of a tetraparental mouse are bred, some of their inbred grandchildren will inherit two copies of the recombinant DNA and, if it was initially inserted by site-specific recombination, no normal alleles for that gene will come along.

When a fully recombinant, tetraparental mouse lives a full life, we can conclude with confidence that the injected gene has been given a chance to carry out all the functions and to display all the meanings it encodes, and we can follow these meanings by following the patterns of expression of the gene and its protein. For example, when the regulatory regions of different brain-specific genes are coupled to a marker gene that produces a blue dye, transgenic mice carrying these constructed genes will show streaks of the dye in their brains, mapping exactly where and when each regulatory region is expressed during the development of the mouse. Or we can simply knock a normal gene out of a transgenic mouse by recombining a silent DNA in place of the EC cell's allele to learn whether, and if so where and when, the missing gene is necessary for a normal life.

In one of the earliest experiments along these lines, EC cells had their p53 genes silenced. The prediction was that in the absence of a functional p53 gene, an embryo's cells would go crazy, making a little tumor instead of a mouse. To everyone's surprise, these transgenic mice developed normally, even when they had absolutely no p53 gene activity in any of their cells. They did succumb to a host of tumors in their middle age, confirming the notion that p53 keeps tumors from growing. But by living for months with no apparent problems, these mice forced a fundamental reassessment of p53's meaning. "Knockout" transgenic mice can also display the contribution of a single gene to even the most complex and poorly understood forms of behavior. For example, scientists have found that knocking out one gene encoding a protein that puts phosphorus on a small set of membrane proteins will yield a transgenic mouse that has lost the capacity for long-term memory.

Transgenic mice descended from tetraparental grandparents are exotic, and making and breeding them requires a rare ability to work equally well with cultured tumor cells, recombinant DNA, and colonies of mice. Wouldn't it be simpler just to inject recombinant DNA directly into the nucleus of a fertilized egg cell? Indeed, if the needle is fine enough and the egg is impaled between fertilization and the first cell division, this is the second way to make transgenic strains of mice. It lacks the finesse of an EC cell's specific recombination, because injected DNAs paste themselves at random into the cell's genome. But when site-specific recombination is not needed, direct injection is an easier way to get a gene into every cell of a mouse.

Implanted in the uterus of another female mouse, the injected fertilized egg grows into a many-celled embryo, and the injected DNA is always copied as if it were a cellular gene. Since the new gene has usually integrated in only one chromosome, it is not yet established as a true-breeding mouse gene. As with tetraparental mice, a DNA-injected mouse founds a new strain by mating to an ordinary mouse. Half the litter will carry the injected transgene; when these offspring are mated brother to sister, a fraction of their offspring will have two alleles of the transgene in each cell. Brother-sister mating of the founder's grandchildren will establish the novel strain, each mouse homozygous for the new gene forevermore.

The gene for SV40 T-antigen was one of the first to be inserted into a mouse this way. As expected, if it was provided with a regulatory region that permitted transcription, the protein caused tumors to appear as the transgenic mice grew up. If the molecular word processor was used to link T-antigen's coding region to a regulatory region for a gene active only in one tissue — a liver-specific gene, for instance — the transgenic mouse would come down with tumors of the liver. Even though every one of its cells had the gene, only liver cells could transcribe it and start T-antigen on its malignant way.

A recombinant gene may also be modified prior to injection so that it can be easily rescued from mouse DNA at a later date. One such transgenic strain of mice provides a particularly elegant way to measure the mutagenic effects of chemicals and radiation: the mouse is given a toxic treatment; then the transgene is recovered from its tissues and assayed directly for accumulated mutations. Mice of this strain are for sale; they are called the Big Blue Transgenic Mouse Mutagenesis Assay System, a big but appropriate name for a mouse with an extra bacterial gene that — once recovered and put back into the right bacteria — will turn a bacterial colony blue. It would be fitting to require that Big Blue transgenic mice be used to measure the mutagenic activity of novel herbicides sprayed on the transgenic herbicide-resistant food plants before the fruits and vegetables could be marketed, but that is not likely, at least until the real cost of a new technology is taken into account more carefully than it is today.

A gene intended for integration into a mouse by EC-cell or direct injection does not have to come only from a virus, a bacterium, or a mouse; transgenic mice have been produced with human alleles that function well enough to compensate for damaged or missing mouse alleles. For instance, the transgenic grandpups of fertilized mouse egg cells injected with a human hemoglobin gene produce functional hemoglobin; when the fertilized egg cell comes from an inbred mouse strain suffering an inherited blood disease, the transgene cures the grandpups' symptoms.

The ability of mouse cells to incorporate human genes is very useful in the study of diseases, especially cancer. The potency of new drugs must be tested directly on tumors, and pharmaceutical firms used to inject tumor cells into mice and then add their drugs, a procedure no more reproducible than the constancy of the tumor cells from experiment to experiment. Since 1988 they have been able to turn to a transgenic strain of mice, patented by Harvard University, that has had a human oncogene — called *ras* — inserted in its genome. Licensed by Harvard, the Du Pont Corporation had some success in marketing this strain, called OncoMouse; Du Pont guaranteed that each untreated OncoMouse would develop a lethal tumor within a few months of birth. Cancer labs all over the world began to use the OncoMouse to test possible drug treatments until Du Pont insisted that investigators who developed products with OncoMouse's help had to pay royalties. At that point, many laboratories simply made their own transgenic strains by injecting a human oncogene into a mouse embryo, bypassing the Harvard–Du Pont product.

Big Blue and OncoMouse are not alone: there are HIV mice as well. No ordinary mouse comes down with AIDS, because HIV is so compulsively specific about growing only in one type of human immune-system cell. But an HIV transgenic mouse's immune cells are damaged from within by the HIV genes in every cell; once HIV is bred into the mouse's genome, no further infection is necessary. These mice, tricked from conception by a molecular Trojan horse, offer a model for studying AIDS that develops the disease's symptoms without ever being exposed to the virus itself.

One transgenic mouse strain rich in nuances shows us how wonderfully alike are the homeobox genes of flies and mammals while providing a new way to study a serious birth defect. It has had one of its homeobox genes — called *hox-1.5* — knocked out. The loss of the second copy of *hox-1.5* produces a profoundly damaged mouse, without a thymus or a parathyroid gland and with a number of defects in the formation of the face, the heart, and the head. Two facts about this poor, sick mouse show the power of transgenic technology to explicate the full meaning of a gene and the rigorous preservation of meaning some genes can display even in extraordinarily different contexts. First, these are precisely the sort of defects caused in the fly by mutations of homeobox genes called *Zerknüllt* and *proboscipidia*, whose sequences are strikingly similar to that of the *hox-1.5* gene. The meaning of these mutations is plain: the helix-turn-helix domain and the other interactive domains of *hox-1.5* proteins have affected the development of embryos in the same way for a much longer time than either the fly or mouse has been around, longer than it has been since the last common ancestor of flies and mice was alive. Second, these mutations suggest that people and mice will hardly differ at all in terms of their ways of reading this gene: the birth defects of a *hox-1.5*-deficient mouse are remarkably like those of an infant born with a congenital deficiency called DiGeorge's syndrome. We do not yet know whether the gene that is mutated in DiGeorge's syndrome is the human version of *hox-1.5* or another gene that shares a developmental pathway with it. Either way, the similarity of phenotype in humans and mice suggests that *hox-1.5* knockout mice will be helpful in understanding the human disease.

The *hox-1.5*-deficient mouse is one of an increasing number of transgenic knockout strains that help elucidate inherited diseases of our own, and new medically important transgenic animals are on the way. A strain of transgenic pigs, for instance, may one day replace blood banks. The pigs express a variant of the human gene for hemoglobin. The variant gene — a sequence created in the lab — makes a hemoglobin that will work as a protein without having to be wrapped in a red blood cell. Hemoglobin can be repeatedly harvested from the transgenic pigs and purified for further experimentation. If this line of work is successful, many transfusions of whole blood may no longer be necessary, and emergency operations may be considerably less risky.

Transgenic mice that mimic Alzheimer's disease may also be at hand, although at least two early reports of such models have been retracted after they could not be repeated. In at least one case, though, a form of the disease did develop in a strain of transgenic mice. It was created by introducing a portion of a human gene that is overexpressed in the dying brain cells of an Alzheimer's victim. The transgenic mice are born healthy, but in time their brain cells fill with tangles of threads of this gene's protein as they become demented. The tangles kill brain cells, leaving remnants that pock the parts of the mouse brain responsible for memory, a terrain quite reminiscent of the brains of Alzheimer's disease victims. This cannot be the whole story, though, because while some cases of Alzheimer's disease run in families, most seem to arise spontaneously — or after an injury — and it is unlikely that these instances of dementia could have all stemmed from any single inherited mutant gene.

One clue that might reconcile the complicated patterns of incidence of Alzheimer's disease with the simple results from transgenic mice is the dementia of Down's syndrome: people who inherit three copies of chromosome 21 instead of the usual two become demented at an early age as their brains come to resemble those of Alzheimer's victims. Most neuronal cells in the brain do not normally divide, but injury can stimulate cell division, and with it the possibility of a disordered mitosis that would put a third copy of chromosome 21 in a new neuronal cell. If that were to happen often enough, a person might spontaneously develop a nonfamilial form of Alzheimer's disease late in life due to the overexpression of a neuron-specific gene from three different copies of chromosome 21, as a Down's syndrome person does at an early age. This way of explaining the data makes the prediction — as yet untested — that familial late-onset Alzheimer's would result from inheriting a mutant allele for any of the many genes that contribute to proper chromosome assortment during mitosis.

~

Though these examples show that we can discover the full meaning of a gene by putting it in a transgenic mouse, transgenic mice will tell us little of what we want to know about the meanings of the human genome. We know there are at least a hundred thousand genes in our genomes, and transgenic mice are not going to be able to tell us about more than a few of them in the next few years, one at a time. Moreover, many traits worthy of study are not the result of one gene's expression but the combined consequence of many genes working together. In principle, we could generate a palette of transgenic strains and then breed them to produce mice that carry any number of different transgenic genes, but this elaborate protocol would go up in cost and complexity very rapidly as the number of human genes involved increased. Besides, a mouse is not as smart as any of us. The parts of our brains that are most interesting, most different, and most obscure to us are assembled and maintained by sets of genes that are not likely to be at work in understandable ways in mice. Some sets of genes, like the ones responsible for the regions of the brain dedicated to language, would probably not even work in transgenic chimpanzees or any other primates, though to a remarkable degree we share genomic coding sequences with them.

Why not transgenic people then? Certainly, there is no technical barrier. The success of in vitro fertilization (IVF) has shown, in passing, that the one-celled human embryo is entirely accessible. IVF is pursued by thousands of couples unable to conceive in the ordinary way; many have had healthy babies by this method. After sperm meets the egg in a dish, the fertilized egg is typically allowed to divide a few times before it is placed in the mother's uterus. For those hours or days, the earliest of human embryos is available for the injection of any DNA sequence one might have made for the purpose. IVF thus presents an undeniable temptation to make a transgenic child.

Most people recoil instinctively from this prospect, since it immediately raises the specter of eugenics and a brave but horrifying new world. But perhaps that is overreaction: what if, for instance, a transgenic child can be born without a congenital disease passed on to her by her parents? Is there a way to use the technology of transgenic development to benefit people and do no harm? In other words, can there be a transgenic medicine consistent with the Hippocratic Oath? The question is neither technical nor conceptual: it is easy to imagine a time when physicians know enough about the genome to say with confidence that one or another sequence would, if properly inserted in a fertilized egg, stop the recurrence of any one of a number of familial diseases. For instance, our medicine already accepts a diagnosis of cystic fibrosis before implantation, and early experiments are under way to replace mutant CFTR protein in the lung cells of cystic fibrosis patients by infecting them with a genetically modified virus carrying a functional copy of the CFTR gene. There is no technical impediment to a combination of these two protocols: inject the CFTR gene into one of the remaining seven cells of an embryo that has been shown to lack a good CFTR allele, and implant the transgenic embryo into a waiting uterus.

The questions that must be answered before such transgenic

medicine — the ultimate in planned parenthood — gets started are social, not scientific. We need to know who is to decide to create the first transgenic embryos. Then — since we can be sure that mistakes will be made — we need to know who is to take responsibility for them. A mistake in this context means, of course, a child born with a defect caused by some step in the experiment. Recently, for example, scientists interested in coloring the hair and eyes of an albino strain of mice transgenically injected the DNA for a missing enzyme that makes black pigment; unexpectedly, they created a strain of transgenic mice whose viscera — heart, stomach, liver, and the like — were all turned around. These mice were unable to live long after birth; careful molecular work showed that the added gene had — while inserting itself in the mouse genome — inadvertently damaged a hitherto unknown gene responsible for maintaining the usual asymmetries of the mouse's internal organs.

Confronting the first, experimental phases of transgenic medicine raises questions that should interest religious leaders, politicians, educators, and parents at least as much as they interest physicians and scientists. The answers we provide — and the nature of the political discourse by which we arrive at them — will determine not so much the medicine of the next century as the manner in which we will live with one another. The issues raised by the potential birth of a failed, experimental, transgenic child are at least as thorny and contentious as the contemporary question of who should decide whether a fetus can be aborted. The passionate and unresolved arguments between "pro-life" advocates, who would forbid abortion as murder, and "pro-choice" defenders, who believe every woman has a right to an abortion on demand, are a foretaste of the disputes that would arise if it were known that any corner of the medical establishment was about to make it possible for a woman to carry a transgenic fetus to term without being able to assure her that the baby would be free of any inadvertent side effects of the DNA injection. Since responsible scientists cannot promise that all experiments will work, I do not see how transgenic medicine can ever be properly undertaken in a democratic society.

Readers of human DNA are not yet at an impasse, but we are surely on the way to one. In a short time we will have transliterated the human genome, played out many parts of it in cells, and gotten some difficult passages translated for us in transgenic animals. Then we will begin to plumb the genome for the passages we want most to know about, the ones that are responsible for our humanity and, if not our souls, then our ability to imagine souls. What we have found about the genome so far suggests that it will be impossible to know ourselves this way through our genomes, but we could easily do great damage to one or more future fellow citizens by trying. As a friend once said to make me think again about performing a particularly seductive experiment, if it isn't worth doing, it isn't worth doing well.

6

BETWEEN PHYSICS AND HISTORY:

THE NEW PARADIGM

OF BIOLOGY

TWENTY YEARS AGO the philosopher Thomas Kuhn suggested that a mature science is always guided by a set of theories, standards, and methods, which he called a paradigm. As he explained, this set of tacit or explicit assumptions determines what is seen as important and what is discarded as trivial or unanswerable, focusing a field and often leading to a rapid consolidation of disparate and confusing problems. But as observations accumulate that do not fit the current paradigm, scientists find themselves constrained by what had been a helpful set of standards, one they cannot discard until they find and agree on another, more useful set. As data are accumulated by scientists in a changing field, the same information may be interpreted in different ways, and old, familiar facts may take on new meanings as well. A temporary crisis of meaning occurs when one scientist sees things in a new way while a colleague has not yet come around. Inevitably, scientists who disagree on the meaning of their data will disagree on what to do next until a new paradigm eventually replaces the old and — by agreement of its practitioners — questions that had been marginal become central to the new agenda of a science, and vice versa.

Centuries ago, for example, the intellectual descendants of priests in Europe, Africa, and Asia — the ancestors of today's cosmologists and astronomers — learned to predict the location of planets and stars with considerable accuracy. In order to explain their observations and make accurate predictions while holding to a paradigm that assumed that the earth lay at the center of our solar system and that heavenly bodies traveled only in perfect circles, these early European astronomers were obliged to invent planetary orbits of increasing complexity, intricate circles within circles. A fifteenth-century Polish astronomer — his equivalent of a Nobel Prize was the latinization of his name, from Kopernik to Copernicus — broke the paradigm by reinterpreting available data under the assumption that the sun, not the earth, was at the center of our part of the universe. With that, the extra circlets of planetary motion were no longer necessary; although they were no less efficient at predicting the motion of the planets, they ceased to be of interest. In recent years biomedical science — and especially the sort of science that places the molecular biology of the human genome at its center — has been approaching just such an impasse, one that may herald a new paradigm.

The molecular biology that built its foundations on the presumptions of physics was attended at its birth by a considerable number of physicists. (Francis Crick and Max Delbrück, for instance, were both physicists before turning to biology.) The basic presumption of classical physics — that our cosmos is governed by mathematically precise laws at all scales, from the inside of an atom to the totality of the universe — had undergone its own wrenching reinterpretation with the explication of quantum mechanics. In the 1920s, new data could only be interpreted by using the ideas that the most fundamental atomic phenomena are based on chance and — even more disturbing — that they are not objectively stable but change according to how they are observed. No longer was it possible to imagine that an atom's position and direction could both be known at once; any certainty about the physical world had to be built up from the statistical probabilities and essential ambiguities of quantum mechanics. The deterministic formulas of classical physics still worked for visible objects, which obeyed mathematical laws with precision, but the positions and movements of the invisible atoms that make up all objects were now lost in a probabilistic blur. While the universe remained determinate, the revised mathematical constructs of physics — more accurately reflecting the way atoms behave — could no longer promise completely predictable events and objective, universal knowledge.

With great insight, the earliest molecular biologists extended this new paradigm of physics to biology: living systems, they reasoned, must contain vast numbers of atoms in order to avoid the indeterminacy of small numbers of atoms, and the information that living systems use to construct and replicate themselves must reside in the specific assembly of large molecules from many atoms. Erwin Schrödinger — the inventor of a formula expressing the eerie new reality that atoms and their constituents were neither waves nor particles, but could be conjured as either by the proper choice of detector — was one of the first to use the new physics to address an important question in biology. In *What Is Life?*, published in 1947, he explained why even the smallest living thing had to be made of many atoms. He also called upon statistical mechanics to cut through a thicket of midcentury genetics and correctly predicted that genes would be made of a large, crystalline, but nonrepeating chemical.

In the second half of this century the adopted paradigm of physics has become remarkably fruitful for biology, but it has not converted biology into a branch of physics. Instead, the early ingenuity of physicists who saw the need to study large molecules has brought biology to a stage of certainty — a belief that all answers exist — quite like the one achieved by physics almost a century ago, just before the discovery of quantum mechanics. This faith has survived undimmed for many decades,* but molecular biology now confronts a new and unpredicted uncertainty, a boundary on our ability to know the final meaning of the genes we study. Where once it seemed clear that everything about the living world was intrinsically knowable, we have found — against our expectations — that when sets of genes combine to become the bodies

---

* As Sidney Brenner put it in 1992: "In biology I have said before that in a sense all the answers exist in nature. All we need is the means to look them up, and that's what the techniques [of molecular biology] give us."

and minds of intrinsically unpredictable people, the complete and final meanings of these sets cannot be fully captured. Single genes may one day be totally understood — although that remains to be seen, given the richness of meaning in genes like T-antigen — but the overall meaning of a genome will not be a predictable sum or product of these separate meanings.

The intrinsic incomprehensibility of our own genome cannot be fitted inside the paradigm, derived from physics, that has kept the search for a complete set of time-independent biological mechanisms at the center of molecular biology. For classical physicists, the dust settled when it became clear that tangible objects behaved predictably, although made of unpredictable atoms. For biologists, the challenge will be to come to terms with the discovery that although individual genes may behave predictably, they do not together form a predictable — let alone completely knowable — genome. Experimentation on human genes, no matter how imaginative, will never give a single, complete meaning to the human genome.

Not all biologists have been under the sway of the classical molecular paradigm. Those studying the causes and consequences of natural selection have no hope and no wish for eternal laws: physicists can predict the next solar eclipse, but no one can predict the next species. Trained as scientists but thinking like historians, biologists studying evolution have always embraced the contingent aspects of current and past life and accepted that we will not come this way again. But even though many students of evolution now use the tools of a molecular biologist, their work has for the most part failed to inform the agenda of molecular biology.

This has begun to change. Many molecular biologists are becoming historians in spite of themselves. As they find more examples of the richness of a true historical record — the many-layered, encrypted meanings of an evolved gene — they are beginning to approach the genomes of individuals and species, not as one would approach molecules alone but as one would approach a library of ancient books documenting the history of life on this planet. And as they learn more about ancient gene families that can play similar — but subtly different — roles in the life of a yeast, a fly, and a person, the historical relatedness of all genomes and the intrinsic unpredictability of their futures have slowly but surely amounted to a new way to see molecular biology itself.

The trend is clear: we can expect to find more and more examples of the richness of a real language in our cells. DNA and protein have grammar and syntax, and we have already come upon typographical errors, double meanings, synonyms, and other subtleties. Future studies will build on the fact that the only assembler and preserver of DNA texts until now — natural selection — has generated a matrix of historical, causal relationships so complex that no code linking natural or synthetic DNA sequences to survival of species will be found, and for this reason natural selection will not be reduced to a set of predictable laws.

Once we set aside the futile search for such laws, we will see that genomes like ours, with their networks of interactions and their great multiplicity of meanings, leave us free to use our imagination as we read them. Seen from the vantage point of this new, historical paradigm for biology, genomes that cannot be fully comprehended as the sum of the separate meanings of their genes are nevertheless enormously exciting, while the sort of genome that was believed to be understood as soon as it was transliterated would have left biology a closed science and a servant of its own technology. But imagination is not enough: every pair of people and every pair of living species — whether peach and frog, elephant and dung beetle, *E. coli* and lobster, or chimpanzee and human — have shared a final common ancestor species at some point. To be sure we have access to the texts themselves, we will need to preserve human individuality, despite many temptations to smooth its rough edges, and to understand the historical details of our descent from common ancestors, a history we share with all other creatures, however dissimilar in appearance.

My generation of molecular biologists will likely continue as before: the indeterminacy of the human genome as a whole enlarges the context, but does not reduce the importance, of understanding any allele within it. When a paradigm is supplanted, Kuhn observed, it is the younger members of a field and those just entering it who are quickest to adapt; a real change in direction will come with the next generation. As my future colleagues now in high school and college stake out their own challenges, they will be able to take for granted the central importance of keeping the earth's entire genetic database alive; of establishing the historical record of life's evolution and elucidating the mechanisms of speciation that link all living things by common ancestry; and of uncovering the deep links between our uniquely human use of language and the genomic language used by our genes. Their goals will be, first, to learn the genomic histories of life on Earth, of our species *Homo sapiens*, of our brain's unique capacity for language and thought, and of languages themselves, and then to understand the ways in which these seemingly disparate histories are actually linked and interdependent.

～

To see these links, it helps to shrink the earth and speed up time by about ten-million-fold each. Now the seven-thousand-mile-diameter earth is about five feet across, not quite as big as our previously magnified SV40 virus. The few-mile-thick layer on the earth's surface that supports all life — from bacteria in the oceans to bacteria in the atmosphere — is a coating no thicker than a dime. Since a billion years of the past collapses into a hundred years of speeded-up time, this miniature earth would have been formed about two thousand years ago. The first living cells would have formed on it almost four hundred years ago; the large creatures with enough bony structure to leave fossils would have lived and died about nine months ago, and the first modern humans — *Homo sapiens* — would have been born about a week ago.

Collapsing perspectives in this way allows us to see that the membrane of living creatures wrapped around our planet is itself a slowly developing, living creature of a sort. Life in this planetary wrapping had a beginning, just as a body begins from a fertilized egg; and from that beginning it has developed as different species — tissues — interact in various ways. Some are shed, others saved; most change, some stay pretty much the same. The growth of cells from a fertilized egg into a living creature is called development; the development of life on

Earth is called evolution. Development and evolution have these things in common, although they work on vastly different dimensional and time scales: they both bring out structures of astounding complexity from the interaction of genes and proteins; they are structured hierarchically; they are time dependent and historically rooted; and, as we are finding out the hard way, neither is totally knowable or predictable.

In some organisms — worms provide a particularly good example — every cell division can be followed and every cell of an adult organism assigned a full genealogy back to the fertilized egg. The totality of genealogies of all cells in an adult organism is called a fate map; the eventual fate of the descendants of every cell in the developing embryo can be read from it. Biologists want a fate map for complex organisms more like ourselves, both to link gene expression to development and to understand more easily the meaning of a given pattern of gene expression in an embryo. We do not have many fate maps yet, but we are confident that they would all look the same in one way: they would all be like a tree, with a trunk from the fertilized egg growing outward through time into many branches and twigs of differentiating tissues and organs.

The fate map of all creatures on Earth looks like the fate map of an embryo. It too begins at some point in time and space — the origin of DNA-and-protein-based life from replicating, catalytic bits of RNA — and grows through almost four billion years (four hundred years in our speeded-up world) into the full complexity of life on Earth today. The fate map of all species alive today is called a phylogenetic tree; it reflects countless specific interactions among species living and dead since life began.

The genomes of all cells in an embryo show a close to perfect resemblance to the genome of their initiating cell, the fertilized egg. Similarly but less perfectly, the genomes of all life on Earth show their descent from a common ancestral sequence. With so much more time for random mutation, DNA movement, and gene duplication, the sequences of species have diverged considerably from one organism to another, but they still show clear signs of their common descent. Proteins, for instance, come in what Ford Doolittle of Dalhousie

University in Canada has called first, second and recent editions. The genes for first-edition proteins are found in all cells; proteins that convert sugars to energy fall in this category. Second-edition proteins are found only in eukaryotes: examples are the histone proteins that wrap DNA for storage in the nucleus and the proteins that confer movement, like actin. Recent editions are found only in plants or in animals but not both: collagen and chlorophyll are examples. There are only about a thousand first-edition genes, each of them very old. Bacteria have one or only a few of each; in their place, the larger chromosomes of cells with nuclei have accumulated extensive families of related descendant genes.

Molecular-genetic fate maps and the phylogenetic tree are hierarchical, because all the branches on each trace back to a single starting point, and because at any level, all twigs on a branch share the attributes of that branch. We can see this in the way we are located on the phylogenetic tree: at the tip of our twiglet we stand alone as a species, *Homo sapiens*. Moving down the twiglet, we come to the twig our twiglet grew from, the genus *Homo*. All other species sharing this genus with us died some tens or hundreds of thousands of years ago. To indicate that the ancestor of all members of the genus *Homo* itself shared an earlier common ancestor with other creatures, we place the ancestor of *Homo* and all its descendant species in the family *hominidae*, the hominids. All other hominids, like all other members of the genus *Homo*, have long since died out. Our closest relatives among the living primates — that is, the ones with whom we share the most recent common ancestor among the extinct primates — are in the superfamily of *hominoidea*, made up of three families: the gibbons, the pongids (orangutans, gorillas, and chimpanzees), and the hominids (us).

We shared an ancestor with the apes, monkeys, and prosimians who did survive, and that shared ancestry puts us with them in the order primates. Primates share body characteristics like live birth, breasts, and hair with a large number of other animals; we indicate this older and broader shared ancestry by placing ourselves in the class *mammalia*, the mammals. Mammals in turn share a backbone with a spinal cord and other anatomical structures, including a common overall body plan, with a host of other animals; these shared characteristics put us in the phylum *chordata*, the chordates. Even earlier, chordates and all other many-celled animals — including the fruit fly, for instance — shared a common many-celled ancestor, and so we and they are all linked in the kingdom metazoa.

Metazoa share nucleated cells with the *metaphyta* or plants, the fungi, and the single-cell protists; creatures in these kingdoms are linked by common origin from the first cell to have a nucleus, which places us on the trunk of the phylogenetic tree in the superkingdom eukaryotes. Finally, we and all other living creatures, whether eukaryotes or prokaryotes without a nucleus, descend from the root of the tree, a common ancestral cell. Although the last of those cells died billions of years ago, we can be confident that the ancestral cell had the attributes that all cells retain to this day: a genome of DNA encoding a set of about a thousand first-edition enzymes to do the work of replication and metabolism, a genetic code, and an RNA-based translation apparatus.

The farther back toward the trunk we go on the phylogenetic tree, the earlier we are in the history of life. Until about six hundred million years ago — when the current phyla diverged — few living things left fossils. Reconstructing earlier chapters in the history of life is strictly a literary enterprise, dependent on the close analysis of conserved DNA sequences. Sequences that are conserved over the widest range of living things are current versions of the most ancient genes. The ribosome genes of our mitochondrial DNA, for example, have been used to confirm the initially controversial hypothesis of Lynn Margulis, of the University of Massachusetts, that symbiogenesis — a series of symbiotic invasions of one primordial cell by another — made possible the evolution of all subsequent eukaryotic life.

According to this hypothesis, more than two billion years ago a primordial purple bacterium set up a stable life inside a primitive eukaryotic cell; in time this bacteria and its descendants lost their freedom to live anywhere but inside another cell, thereby becoming the symbiotic ancestors of the mito-

chondria inside almost all eukaryotic cells alive today.* A
subsequent seduction of eukaryotic cells by bacteria — this
time blue-green ones, which could turn sunlight into sugars —
gave the ancestor of today's green plants the beginnings of
what would become chloroplasts. We can recover signs of
these long-term symbiotic events — palimpsests like the Old
Testament phrase in the line from James — because chloroplasts and mitochondria still retain genomic circles of DNA
from the days when their ancestors were free-living bacteria,
and these genomes still carry genes for their own ribosomes.
Once the sequences of these genes could be compared with the
sequences of ribosomal genes from the cells themselves and
from various strains of bacteria, it quickly became evident that
ribosomes of chloroplasts and mitochondria are not similar
to the genes of any eukaryotic ribosomes. Rather, they most
closely resemble the genes of certain bacteria that grow in hot,
sulfurous springs, an environment evocative of one of life's
more auspicious homes a few billion years ago.

The efforts to construct molecular-genetic fate maps — the
trees of gene-regulatory pathways that accompany differentiation and development in a single plant or animal — and to
recreate the phylogenetic tree of life on Earth are related in a
simple way. Disparities of size and duration — individual organisms are smaller than the planet; individual lifetimes are
shorter than the lifetimes of the species — have until recently
kept these two fields from informing each other's efforts, but
to obtain the fullest possible meaning of the individual genes
they study, molecular biologists will have to know a gene's
history and the roles played by homologous genes in other
organisms. The Harvard geneticist Richard Lewontin caught
the cusp of this in 1982, just as the tools of the molecular word
processor were in their earliest stages of development: "The
great evolutionist Theodosius Dobzhansky wrote that 'nothing
in biology makes sense except in the light of evolution.' But

---

* An exception worth mentioning is the intestinal parasite *Giardia lamblia*, which has no mitochondria at all. Its nuclear genome diverges from all
other eukaryotic genomes; *Giardia lamblia* is perhaps the sole descendant of
a eukaryotic cell line that lived before the mitochondrial invasion.

we must add that 'nothing in human evolution makes sense except in the light of history.'"

A new, historical approach has already made itself felt in molecular biology: every newly sequenced piece of a genome is "looked up" by computer programs that use the fact of common descent to search for sequence homologies in a growing data bank of hundreds of millions of base pairs from hundreds of species. Coding sequences that have no homology with any known gene are increasingly rare, and it is possible that we may learn the vocabulary of regulatory domains before too long. In filling out these time-dependent, historical maps — the DNA and protein sequence analysis of homologies among DNAs, the phylogenetic reconstruction of relationships among all living creatures, and the molecular biology of development — some molecular biologists are already contributing to a more complete, historically informed biology.

At the same time, the mechanistic lines of inquiry that until recently defined molecular biology have increasingly become dependent on their utility and profitability rather than their intrinsic interest. While laboratories isolating genes associated with human disease or crop development remain magnetic attractors of private and government funds, the five hundredth such human gene hardly generates the buzz that heralded the first five. In place of the simple cloning and sequencing of a human gene, critical experiments now revolve around the use of transgenic mice to reveal the fullest possible meaning of the gene in the richest available context, and the centrality of the recovered gene itself has begun to be displaced by the possibility of creating — writing — a new version of it in our word processor, and then drawing out our own version's meaning in the transgenic context.

∼

To see the world as a sphere covered in a thin skin of tissue called life, with a particular history that will not happen again, is to bring the variable of time to molecular biology and thus allow genomes, their encoded proteins, and the networks of regulation that bring these proteins into the world to be seen as natural creations sharing two properties with literature, art, and science itself: knowable pasts and unknowable futures. In the case of our own genome, time and evolution have led to the unbounded and unpredictable attributes of consciousness and language. Historically based molecular biology will want to understand the genomic contribution to these two uniquely human attributes. Acquiring this knowledge begins with a fuller understanding of the origin and history of our species, so let us return to the phylogenetic tree to see how we compare with our nearest neighbors.

The first hallmarks of a future that would include us are in ancestral mammals who lived at least two hundred million years ago. From that stock, mammals diversified — sometimes slowly, sometimes quickly, especially after the cataclysmic death of the dinosaurs sixty-five million years ago — into a set of about four thousand living species, different enough from each other to be placed in no fewer than fifteen different orders, including our own, primates. Recall that among primates, we share the most characteristics with other members of the hominoid superfamily of gibbons, orangutans, gorillas, and chimps. Our genomes and those of the great apes — particularly the African forms — are very similar, with more than 98 percent sequence homology between the genomes of chimp and human and not much less between human and gorilla or chimp and gorilla.

The last common ancestor of the living hominoids died off no later than ten million years ago, and for much of that time none of the hominoids were particularly human. About eight million years later — two million years ago — the first *Homo* species whose fossils we have found, *Homo habilis*, appeared and lived among other, smaller-brained cousins, the Australopithecenes. *Homo habilis* made tools, was about three feet high, and had a brain volume of about twenty-three ounces. In rather short order a bigger species, *Homo erectus*, appeared in Africa, in time supplanting *Homo habilis*, and spread to parts of Europe and Asia, living from about 1.8 million years ago until about four hundred thousand years ago. By then various strains of *Homo erectus* had grown to about six feet tall and had up to thirty-three-ounce brains.

The pace of change was accelerating: about three hundred thousand years ago, even the biggest and brainiest *H. erectus* had died off, and in their place we find the fossils of the first of our own species, archaic *Homo sapiens*. These early humans, with brains at least as large as ours at forty-five to fifty ounces, left a cultural heritage that includes the wonderful grave at Shanidar in Iran, in which an archaic *Homo sapiens* of the Neanderthal type was lovingly buried with flowers. Anatomically modern *Homo sapiens* first appeared in eastern Africa about a hundred thousand to a hundred and twenty-five thousand years ago and began migrating soon thereafter. Europe, Asia, and Africa saw many millennia of joint habitation by archaic and modern humans, but the Neanderthal people of Europe died off about forty thousand years ago, and we have been the lone members of the genus *Homo* ever since.

The molecular biological investigation that would help us understand many of the open questions in this history has only just begun. We do not know, for example, whether modern and archaic *Homo sapiens* ever interbred, nor in particular whether current Europeans carry any DNA sequences inherited from the Neanderthals. There is evidence, however, that modern *Homo sapiens* lived for some time in Africa before migrating north, west, and east to every continent beginning about a hundred and fifty thousand years ago, quickly establishing all races of humankind in the process. For example, certain stretches of mitochondrial DNA are more diverse in Africans than in people of any other continent, suggesting that their ancestors — who might have lived in Africa for some time before their great migration — had more time to accumulate allelic variations than did the small groups of migrating African ancestors who became today's Asians and Europeans. But the evidence is not conclusive, and some paleontological evidence supports the possibility that Europe and Asia — both possibly populated for a million or more years by descendants of the early migration of *Homo erectus* from Africa — were also the home of their own versions of archaic *Homo sapiens* before the second migration from Africa took place. In that case, some scientists have argued, the modern peoples of Europe and Asia might be the descendants, not only of the exodus of African *Homo sapiens*, but of separate, locally evolved versions of *Homo sapiens* as well. Molecular homologies fall on the side of the hypothesis that we are all the descendants of the second, more recent African exodus, but only barely, and much more work needs to be done to reconcile molecular with paleontological data.

Over the sweep of the thousand centuries since we began to establish ourselves on all the continents, modern *Homo sapiens* has been a remarkably promiscuous species. Many of our genes may exist in any of a number of different but functional alleles, and almost all normal alleles can be found — at various frequencies — in human chromosomes all over the planet. We have only begun to characterize the incidence of various alleles in genomes from representatives of the thousands of ethnic groups, but even this earliest work reveals that the genomes of today's Europeans are a hybrid intermediate between Asian and African ancestors. This could not be so unless mating occurred frequently among the ancestors of currently isolated groups of people. The genetic diversity between any two ethnic groups — quantified by measuring allelic frequencies — is usually small, relative to the diversity between individuals in one group. The result is clear: when it comes to our species, you really can't judge a genetic book by its cover. Molecular-historical biology has therefore already told us something important about our behavior as a species: despite our tendency to live or die by the dream that it might be so, none of us has come from a pure ethnic or racial stock.*

~

In his private notebooks Darwin wrote, "Why is thought — being a secretion of the brain — more wonderful than gravity a property of matter[?] It is our arrogance, our admiration of ourselves." Perhaps, but this secretion has made it possible for a DNA-based life form to understand that it is constructed by a text written in an ancient language, and surely there is nothing more appropriate for students of DNA to learn than where in the genome the ability to comprehend any language resides.

---

* In any case, race is a cultural, not a biological, term when it is applied to our species. For instance, as a consequence of the period when African people were bought, sold, and bred as property, "white" people in the southern United States today have a higher frequency of alleles associated with people from Africa than do "white" Northerners or "white" Canadians.

Languages, tribes, and races are all relatively recent developments in our forty thousand years of uncontested survival as
the only hominid. Because the phylogenetic tree allows us to
use comparative anatomy to tell us about our past, we can see
how our survival may have depended on a new use of the lungs
and larynx for speech. The mouth of all primates feeds into
two tubes: one — which is joined by a path through the nose
— goes to the lungs; the other goes to the stomach. In other
primates, the epiglottis keeps the nose-lung pipe open while
directing the contents of the mouth to the stomach, allowing
them to eat and breathe at the same time. In humans, the
position and large size of the larynx — essential for speech —
prevent the epiglottis from separating the two pipes, so we can
choke if we eat and breathe at the same time. Fossil evidence
suggests that archaic *Homo sapiens* had the usual primate
epiglottis; it appears unlikely that the throat of a Neanderthal
could form a complex string of sounds. Our larynx is uniquely
placed to intervene in the flow of air from the lungs. Apparently the risk of choking was overridden by the benefits of
language when we shared the continents with archaic *Homo
sapiens*.

What in our genomes gives us language? When phylogenetic
relatedness is assessed by form, the degree of similarity between individuals or fossils can be measured by counting the
number of shared attributes, usually called characters. Molecular biology has had little to contribute to the notion of
characters; they may have a wide range of molecular complexity. A single base-pair difference may count as a character in
one study, whereas another study will consider the difference
between chimps who lack spoken language and humans who
have it to be one character. Since about four tenths of all the
messenger RNAs made from the human genome are made
only in the human brain, it is possible that a character such as
the ability to form languages has the molecular complexity of
tens of thousands of genes.

At least some of the human genes involved in creating language must be active in the assembly and operations of a small
sector of the human brain named for the scientist Paul Broca;
structures in this region are closely associated with the capacity to speak and understand a language and next to structures

involved in repetitive motions of the arms and fingers. Since we know that complex stone tools of the Achulean type were manufactured by *Homo erectus* as early as 1.4 million years ago, an appealing model for the origin of languages is pre-adaptive; that is, our facility for language may be the result of an evolved use of neural circuits selected in *Homo erectus* or its ancestors for their role in repetitive motions with tools.

Languages, as MIT's professor of linguistics Noam Chomsky was the first to show, are organized hierarchically, and comparative studies of the brains and behaviors of living primates support the notion that the Broca region of our last common ancestor directed complex, hierarchical manipulations of objects, a function it still serves in chimpanzees and in humans until about age two. Patricia Greenfield of the University of California has connected Chomsky's observation with data drawn from observing primate behavior to suggest that the function of the primate Broca region has always been to assemble hierarchies: first, hierarchies of manipulation for the body's hands; later, and only secondarily, hierarchies of linguistic structures for the organs of speech and hearing.

Most primates besides ourselves and the Old World apes have Broca regions that function like those of a one-year-old human child. In mature humans, chimps, and gorillas, the nerve cells of the Broca region are connected to areas of the brain that drive the hands, mouth, and tongue, permitting adult chimps and gorillas to make and use single tools. The Broca region in our brain is bigger than a chimp's, and it also makes connections to regions at the front of our brain that are lacking in chimps and other primates. From age two on, human brains develop neural circuits in and around the Broca region that confer the ability to assemble both complex languages and complex combinations of objects; chimpanzees lack this second cycle of postnatal development. Grown chimps, with Broca regions most like ours, can make a simple subassembly by combining two tools into a third. But only humans use tools to make a new tool and then use that in another assembly, in ever-deepening hierarchies of complexity. And while other primates have a rich vocabulary of purposeful sounds, only human languages organize words hierarchically, and only humans assemble words hierarchically to make a language, using words to make clauses, clauses to make sentences, and sentences to make arguments.

Persons unable to make full use of the human capacity for spoken language can use their Broca regions, their capacity for language, and their hands to communicate in the rich language of Sign. Deaf children raised by parents who use Sign develop their language skills through a formative period of the hand equivalent of babbling, and when Sign is used by a grown person, the Broca region — as well as the regions normally given over to hand movements — is activated in the brain. Small lesions in the human Broca region, though, can cause the loss of both hierarchical capacities: a man with such a lesion can assemble strings of words but no clauses or sentences; he will be able to put blocks in a row, but not to assemble structures from them.

Even as we begin to understand the biology of thought and language, we must acknowledge how little we know about the genomic contribution to consciousness. The frontal regions of our brain, in which abstract notions are processed and which through their connections to the Broca region drive our language skills, develop from a segment laid down in early embryonic development by a member of a family of genes containing homeoboxes. But which homeobox genes are most closely associated with the assembly of the frontal regions of the brain in humans? And did the duplication of a homeobox gene at the "head" of a family of genes, in an ancestral primate a few million years ago, set a primate line on the path to language, knowledge, and thought? Which genes are activated in the Broca region of the human brain, and the regions it feeds, as a two-year-old acquires grammatical language skills? We don't know the answers to any of these questions; it will be the task of a new generation of molecular neurobiologists, versed in the historical context of hominid evolution and the comparative anatomy and genetics of the primates, to search them out.

～

A developing connection between genetics and linguistics gives rise to another, if less likely, series of questions for the new biology. Our myriad languages divide us, even as our capacity for language defines our common humanity. From the global vantage point we are a fractured species, separated by language into hundreds of groups. On every continent many — although not all — languages are wrapped in flags to signify that they have been elevated to the formal autonomy of nationhood. Given the shared power of language, why has our species chosen to separate itself into mutually uncomprehending tribes, states, nations, kingdoms, and empires? How did languages develop among us, separating us this way? Have languages affected our genomes? These hardly seem like questions for classical molecular biology, but if we want to know the history and the full meaning of the many alleles spread among our species, we will very likely need to know the history of the development of our languages from their beginnings. To begin, we might ask: why should the genomes of people speaking different languages be different?

The words of a particular language are arbitrary; it is only necessary that all who speak it agree on their meanings. Let us assume that all living people are the descendants of ancestors who lived in Africa for some long period and that these ancestors used the anatomical and mental capacities of *Homo sapiens* to speak a language. As these ancestors migrated outward to populate Turkey, the Middle East, Europe, Asia, and later Polynesia, Australia, and the Americas, physical separation would have allowed "speciation" of languages: the language spoken in each newly reached, isolated part of the world would have been free to change in ways that might easily render it incomprehensible to the people occupying a different range or continent. Dispersion and lack of contact with other people would therefore be accompanied by the evolution of different languages. In their search for the hypothetical tree of ancestry for languages, linguists who analyze the degree of homology of two languages by comparison of phonemes, words, and grammars have already found evidence for the evolution of languages from common ancestors. As with the phylogenetic tree, however, the further back the splitting, the more difficult to reconstruct the ancestral tongue.

The languages of Europe have been particularly carefully analyzed, and the results suggest that the people of Europe continued to fragment and refragment into noncommunicat-

ing groups well after the hunter-gatherer tribes that first populated the continent were supplanted by farmers. English, for
example, is one of about a hundred languages that have split
from the language — called Proto-Indo-European — spoken by
the first modern humans to arrive on the European continent.
As Proto-Europeans spread west about eight thousand years
ago from what is now Turkey to Greece, then north to what
are now the Slavic and German-speaking lands, then west and
south again to what are now the Latin countries, their language split into mutually incomprehensible dialects over and
over again.

The first offspring languages are — like the progenitors of
phylogenetic orders — long since dead: Anatolian, Aryano-
Greco-Armenian, Celto-Italo-Tocharian, and Balto-Slavo-Ger-
manic. Each became the parent of many languages. Balto-
Slavo-Germanic, for instance, was replaced in various parts of
northern Europe by Balto-Slavic, Northern Germanic, and
Western Germanic, three dead languages that kept the ancient
peoples of northern Europe divided into mutually uncomprehending regions. More recently, the territory of Western Germanic speakers divided further, into regions speaking English,
Flemish, Dutch, Low German, and High German; one of these
regions later acquired the trappings of a nation-state. In two of
the most recent bifurcations, High German and Yiddish split
about a thousand years ago, Dutch and Afrikaans about three
hundred years ago. In all these examples, and in all the other
cases we might examine anywhere on the planet, the appearance of new languages signaled the separation of a group of
people into mutually uncomprehending subgroups, a process
that has gone on since our emergence as a species.

Proto-Indo-European is just one branch — and not even a
major one — on the tree of languages. Just as our genomes
contain many genes with domains derived from a small number of ancient, first-edition genes whose initial functions were
too important ever to be lost by any living cell, so can some of
our words still be traced back to the speakers of our species'
first language, called Proto-World. In both cases the principle
of conservation is the same: the sequences — genes, domains,
or words — that are most widely distributed among today's
species or languages are the ones likely to have existed at the

earliest stages in the formation of the evolutionary tree. For example, the words *haku* and *hita* in Proto-World changed hardly at all to *haku* and *-ita* in Nostratic, the precursor of Proto-Indo-European, then became *hakw* and *hed* in Proto-Indo-European, *aqua* and *edere* in Latin, *wazzan* and *ezzan* in Old German, and "water" and "eat" in English. At the same time but in another part of the world, in Amerind — the root language of many indigenous North and South American peoples — the words again hardly changed from Proto-World, becoming *hakw* and *hit-*.

Once groups of early people separated completely from one another by migration through continents devoid of other humans, neither their languages nor their alleles could be exchanged, so allele frequencies as well as languages were likely to diverge. When Luigi Cavalli-Sforza of Stanford University examined the amino acid and base-pair sequence differences for a set of proteins in representatives of human cultures from all over the planet, he found that people whose languages are known to have separated more recently from a common ancestral language have fewer sequence differences in their alleles than people whose languages have been apart for a very long time.

The initial appearance of separate languages can be explained as the result of the physical isolation of separate migrating bands of hunters and farmers, but why did languages and genomes continue to diverge once people began to share the same territories? When a group of people live in the same region and intermarry over a long period, they will share the same allele frequencies for multi-allelic genes; this sharing is called gene flow. When different allele frequencies appear in a population — whether by random drift or, as with sickle-cell anemia, in an adaptive response to a local change in the environment — they can be maintained only if gene flow is very limited. The link between our languages and our genomes turns out to be simple: language differences are a powerful constraint on the selection of a spouse and therefore a barrier to gene flow.

Because languages keep groups of people in genetic isolation from one another, separate languages preserve local concentrations of alleles that contribute to differences in appearance.

Once this occurs, language and appearance mutually reinforce a positive feedback loop that drives our species into an ever larger number of groups of ever more different-looking people. Not even nations, with all their powers to set boundaries and wage war, are as effective at keeping people apart as languages; nationalism based on language is therefore a particularly powerful way to enforce isolation, even though xenophobia may be a common consequence.

Robert Sokal, at the State University of New York in Stony Brook, has shown how efficient languages are at preserving differences in alleles. He examined the genomes of people from all over Europe, measuring the frequencies of 63 alleles of 19 genes at more than three thousand locales. He discovered that in 33 places in Europe, allele frequencies changed very sharply across short distances, which he called boundaries. Some 22 boundaries were mapped onto barriers that might be expected to keep gene flow down by simple physical isolation: 18 were oceanic and 4 were mountainous. But the most striking barrier was neither rock nor water but language: 9 allele boundaries overlaid no physical boundary but only a linguistic one, 31 of the 33 boundaries separated linguistic families or dialects, and 27 were zones of contact of ethnic groups that had originated far from one another. For example, an allelic boundary in Iceland traveled along terrain separating the regions of Iceland settled by immigrants from Norway and Ireland, a migration that ended a thousand years ago. Unlike the preservation of the sickle-cell allele in malarial regions, such allelic boundaries cannot be the result of adaptation to a specific environment. Instead, they provide strong evidence for the inherent capacity of language differences to preserve and widen genetic differences by inhibiting gene flow.

Studies on languages and alleles have brought us full circle: the human capacity for many languages has had consequences for the genome that confers it. The future development of studies on the genomic contribution to the capacity for language, and the linguistic contribution to genomic differences, will make this branch of molecular biology into a behavioral science. But at the same time, molecular techniques have given us the capacity to be far more intrusive than any other branch of behavioral science: we also have the capacity to change alleles in human DNA by direct intervention, adding a fateful, prognostic voice to the language-intensive but otherwise free choice of two people to have a child.

~

The years that end and begin millennia are notable numbers, deep reminders of our temporary place in the long passage of time. They make us take notice of where we are, the way we do when all the little wheels on our car's odometer turn, as 19,999 becomes 20,000 miles. At the end of the last millennium, the world that kept measure by a Christian calendar awaited apocalypse and the end of days. As this millennium draws to a close on a planet of many religions but — from convenience if not faith — one calendar, a sense of foreboding has reappeared at a critical time in the study of three of the planet's most interesting histories — of life, of *Homo sapiens*, and of language. Since their basic texts are genomic, all three histories will remain obscure until molecular biologists develop their skills at genomic analysis. But as we approach the third millennium, it is apparent that our species, fragmented among nations, has done serious, long-term damage to the earth's capacity to support life itself, so portions of the basic texts carrying these three histories are at risk of being lost before we have even opened them.

As Edward O. Wilson of Harvard details in *The Diversity of Life*, the loss of tropical rain forest in the 1980s has meant the loss of about 1 in 200 species every year. Though precise estimates are difficult to get, it seems clear that each day — as about a quarter of a million new humans join the five billion of us already here — at least a hundred of Earth's five million to fifty million species disappear.* In the last decades of this millennium, a political movement to halt and reverse ecologi-

---

* The broad range in estimates of the number of different species tells us how much work remains to be done before we can claim to have even the outlines of a record of life's diversity. While forms visible to the eye have been catalogued for centuries, new species still remain to be discovered in remote areas, and the terrain of microscopic life remains largely unmapped: a spadeful of soil taken anywhere is almost certain to contain unreported species of microbes.

cal degradation and the loss of species has developed in response to these facts, as people in many countries of the world — speaking many languages, governed in many ways, and worshiping many gods — begin to see the need to think and act as members of one among many species in order not to ruin our common planetary home.

To sectors of this environmental movement, all science is at best a mixed blessing. In their terms, the planet's troubles arose in the first place as one branch of science after another lost sight of its supposed reason for being — the careful elucidation of nature's ways — to serve instead a perverse variant of Jeremy Bentham's utilitarianism, the production of greater piles of goods for greater numbers of people. The result of one final century of belching smokestacks and burgeoning populations, they argue, is a planet that shows signs of serious discomfort with us: a burned-out ozone layer, a blanket of warming carbon dioxide changing our climate, poisonous metals and bits of our Styrofoam coffee cups in the deepest oceans, and so forth. This view of the future can be carried too far, until it becomes a modern version of the blood and soil romanticism of the last century, a fundamentalist revulsion with science and technology that portrays our planet as a single, living organism so glutted and clogged with the outpourings of our technology as to be infected, rather than populated, by us. But environmental activists who have taken the trouble to study the sciences of ecology and evolution and who have at least an acquaintance with their biological, geological, and chemical underpinnings see that while we humans are a problem for the planet, all of us (even the scientists) belong here too. These environmentalists know that our species is but one of many, with no more right to a place on the planet than any other, but certainly no less; and they call on scientists to protect our common and only home in the ways each of us knows best.

The classic molecular biology that studies the mechanisms of the action of specific genes and proteins will not cease — nor will it necessarily be overrun or supplanted — because some molecular biologists see in the richness of DNA texts a new reason to speak out in defense of the preservation of species diversity. But the longer molecular biology clings to its paradigm of physics and keeps itself aloof from matters of history, the more likely it will be that the loss of genetic variation — in food crops, in wild animals, in the millions of unrecorded species, and in our own species — will degrade and destroy just those undiscovered DNA sequences that might have led to a more complete comprehension of our own genes and genomes.

The loss of species by deforestation and other human activities is as pivotal a threat to new molecular biology as the pillaging of Alexandria's library or the burning of books by Hitler's university students were to the enterprise of history itself. As we enter the next millennium, historically grounded molecular biology and scientifically informed environmentalism will find their common ground — if both are alert and flexible enough to grasp the opportunity — in a shared view of the world as a single home for us all and of us all as part of a single human family.

~

Once molecular biologists accept the central importance of preserving and studying the diversity of genomes — some because they are interested in the new questions to be asked of these historical texts; others because human genomes are the only source of nature's experiments on human genes — they will see the need to stem the loss of indigenous peoples and languages. Whether an indigenous people become victims of war or economic casualties or they themselves choose to join a larger society, once they are lost, a portion of the human genome's complete library of drafts — some novel assortments of alleles, perhaps even some novel alleles as well — can never be recovered. While the Human Genome Project labors to harvest a single allelic sequence for one consensus human genome, others are more interested in the diversity of our species' genomes than in acquiring any one version of it. Luigi Cavalli-Sforza, Kenneth Kidd of Yale, and the late Alan Wilson of the University of California informally began what they call the Human Genomes Project a few years ago, a far-flung effort to collect blood samples and DNA from members of vanishing indigenous peoples around the world. Their work has cost very little and gained hardly any of the visibility of the Human

Genome Project, but in time their archive will be appreciated as the first international attempt to preserve our genomic inheritance.

The genomic archive at Yale is just the beginning: a new generation of molecular historians will see the attraction of collaborating with environmentalists in the establishment of a fundamental database, a planetary encyclopedia of genomes. This encyclopedia would be a collection of genomes — prepared as DNA libraries, referenced by PCR — of representative members of the diverse species that live above, on, and in the earth's continents and oceans. Many countries already have museums of natural history, private and public institutions charged with assembling and maintaining collections of creatures great and small. With PCR as a magnifier of bits and pieces of DNA, and with homology between living and dead species as a guide to the synthesis of proper primers, museum collections also offer us a chance to recover portions of genomes from extinct species. DNA sequences have been recovered from the feathers and skin and seeds of dried, pickled, and stuffed specimens of now-extinct species, but PCR — which can amplify short stretches of mitochondrial DNA from insects trapped in fossilized amber a hundred million years ago — has barely been brought to most collections.

While the bodies and tissues of specimens often cannot be easily moved from one museum to another, DNA libraries are as portable as CDs, and transliterated sequences are as light as bytes. Molecular biology makes its home today in universities, medical centers, government laboratories, and private companies. As museums change the way they gather, maintain, characterize, and share their collections, they will join with and perhaps one day regain their preeminence among the venues for the humane but scientific study of life.

~

What lies ahead for molecular biology? It would be paradoxical to have argued that the biological future is not predictable, only to predict the future of biology. But extrapolations from current research can serve to sharpen the issues that all of us — not just scientists, but also citizens, voters, parents, and children — will have to face. Molecular biology, like any other science, has been guided by little more than the tastes and drives of its practitioners. A new, broader context will take hold only if the best of today's scientists see the historical paradigm as a source of new questions worthy of their attention. But in this period of adjustment — between paradigms, as it were — molecular biologists have a rare chance to think again of the possible consequences of their work before they go too far up an expensive blind alley or down a road that leaves the rest of us facing a mess we neither asked for nor know how to control.

Consider, for instance, the kind of information that might come out of a merger of DNA analysis with twin studies. Scientists who compare identical with fraternal twins have shown that the inherited component of any complicated human difference — in height, say, or musical ability, or sexual preference, or intelligence — is usually the consequence of allelic differences in many genes. Until recently, it was impractical to untangle such complex patterns of inheritance. But with the power of PCR and with increasingly facile computer programs, it will soon be possible to identify and analyze the particular alleles that all members of a family carry for hundreds of different genes and DNA marker sequences. As researchers apply the techniques of DNA analysis to the genomes of identical twins and their families, we have reason to expect they will extend the work of scientists like Dean Hamer to discover sets of DNA sequences that correlate with just the attributes listed above.

Recall that Craig Venter has obtained thousands of partial sequences from messenger RNAs expressed specifically by portions of our brains. Venter's many sequences make excellent and appropriate probes for these studies, and no technological barrier would prevent them from eventually being applied in IVF preimplantation analysis as well. Here is how Nick Martin, a scientist engaged in these studies in Brisbane, Australia, sees the immediate future, in a recent interview in *Science* magazine:

> Martin believes the same approach could be used with genes for IQ and personality. "I can't wait to get into it," he says.

And although a practical application for such research is now a "pie in the sky," says Martin, it is exactly the kind of work that could lead to such *Brave New World* activities as measuring the precise genetic makeup of children so that the teaching environment could be tailored to shape their development efficiently.

By the light of such Panglossian eagerness, the importance of the new paradigm in biology is apparent: seeing all the living world's DNA as a set of historically related texts is less hazardous — as well as more accurate — than seeing it as a vastly complicated set of molecules from which laws may be derived, personality traits decoded, or new ways found to "measure the precise genetic makeup of children." No serious historian expects the historical record to yield rules that predict the behavior of a person or the course of history, but biology has all too often provided scientific cover for those who would write such rules into laws governing our behavior, even though every claim to have found such laws in science has turned out to be premature, and some have also provided excuses for terrible, lethal behavior.

My generation of molecular biologists has already come upon the temptations and risks of molecular eugenics, embodied by the development of technologies for intervention in human reproduction on a mass scale. Here, for example, is a description of one component of the procedures used in a 1989 study published in the prestigious British medical journal *Lancet*, "Biopsy of Human Preimplantation Embryos and Sexing by DNA Amplification":

After approval of the project by the ethics committee of the Royal Postgraduate Medical School and the Voluntary Licensing Authority for Human in Vitro Fertilization and Embryology, patients were approached at least one month before proposed IVF for tubal infertility to ascertain whether they would consider donating "surplus" embryos, if available as a consequence of their treatment, for studies aimed at better diagnosis of genetic disease.

That is, women waiting for IVF were asked to give over extra eggs for experimental fertilization, and the resulting embryos were then used to establish that PCR could differentiate male from female embryos. This study was "aimed at better diagnosis of genetic disease," only because the authors intended to use their procedure to identify — not implant — the IVF boys of mothers carrying X-linked mutations.

As this example suggests, we are only just beginning to see what may lie ahead: the direct, eugenic selection permitted by IVF (whether negative selection through assay for problematic alleles or positive selection through transgenic IVF); the development of a molecular pharmacology for controlling human sexuality; or the allelic fingerprinting of everyone at birth for purposes related to bearing or raising children. All these procedures will undoubtedly be proposed with great enthusiasm in the coming years, but once they are seen in the light of the human genome's global impenetrability, these dead ends of an old paradigm will appear as wrong, and as dangerous, and as the establishment of separate culture vessels for alphas and betas in the *Brave New World* of Huxley or the numbering of people by a tattoo on their arms in the real world of German concentration camps.

The genome's indeterminacy provides us with an appealing alternative: to see DNA as literature. One of the best definitions of literature comes from the Italian novelist Italo Calvino; in his *Six Memos for the Next Millennium*, Calvino tried to characterize the essence of great literature. His prescription for the literature of the twenty-first century included five qualities: lightness, quickness, exactitude, visibility, and multiplicity. As it happens, the genome of a person has all of these qualities in abundance. It is light because it is as small as the rigors of natural selection permit. It is quick because it must be: the life of a cell is short, and the entire text of the genome must be copied in a few hours; in the case of an embryo, the genomic text must make a person ready for the world — starting from a single cell — in only a few months. It is exact because its base sequences and the proteins they encode create the specificity of surfaces that gives living things their distinctive complexity and efficiency in a disordered universe. It is visible because cells, the genome's readers, assemble into a living thing from its instructions. But above all, the human genome is multiple. We are different from one another, and this allows the DNA texts within us to carry the infinite multiplicity of possibility in human character and, most especially, in the hopes we have for our children.

Calvino himself saw a necessary connection between science and literature. As he wrote in *Six Memos*:

> Only if poets and writers set themselves tasks that no one else dares imagine will literature continue to have a function. Since science has begun to distrust general explanations and solutions that are not sectorial and specialized, the grand challenge for literature is to be capable of weaving together the various branches of knowledge, the various "codes," into a manifold and multifaceted vision of the world.

In DNA, nature has created a manifest and multifaceted text. Once we finally see that all genomes are a form of literature, we will be able to approach them properly, as a library of the most ancient, precious, and deeply important books. Only then can the new biology be born.

CONCLUSION

LESS THAN A DECADE after the ashes of the Second World War had cooled, James Watson and Francis Crick were making tin and paper models of DNA and uncovering the set of laws that govern all inheritance. Since then we have come to understand that DNA is a chemical text that instructs our bodies in all their operations while copying itself so faithfully that these instructions can be passed from generation to generation, enabling life to persist on our planet. These discoveries have revolutionized biology, and the study of genetics is now so complex and ambitious that it has spilled beyond the boundaries of science: human chromosomes, and the information carried by their DNA, will increasingly guide and perhaps even direct our politics as well as our research in the next century. Meanwhile, some of the deepest assumptions of a free society — each of us is an individual; each of us has a private life; we are all equal under the law — have once again been called into question by the work of biologists. The risks posed by investigations into the workings of the human genome come from nineteenth-century notions of eternal progress through science, this time linked to a new and extraordinary power to change inheritance through the reordering of DNA.

Earlier in this century, physicists were lucky enough to have two decades in which to recast the paradigm of their field and then pursue their discoveries while still inside a cocoon of ostensibly pure, objective science, above human frailty, beyond petty politics, motivated by curiosity and intellectual adventure alone. Though the context of their work changed, they managed to preserve the context of their profession. That they ultimately did not live up to their own professional standards of distance and objectivity — the atomic bomb was as much the work of American physicists as were the theories that allowed them to design it; the expedient anti-Semitism of physicists in Germany canceled the lead German physics had enjoyed in the decades before the war — did not keep them from passing those standards on to us. The idea that scientists ought to work at a remove from the concerns of common humanity remains powerful indeed.

Molecular biologists ought not to repeat their mistake. Somehow, by a grace I do not understand, we have not thrown the switch of nuclear war a second time in the past half century. In this period of grace we have developed the capacity to read, to edit, to rewrite, and perhaps to begin to understand our DNA texts. If we can give ourselves the time, our genomes will together teach us that we are all profoundly related, that we are all one human family. Abandoning hope of final closure on the language meaning of the human genome, future molecular biologists will be free to be critics and historians, expert at multiple interpretations, even able to go back and forth between the white lab coat and the jacket of a critical scholar or the blue suit of a public servant. My colleagues have never had a better opportunity to become engaged in the political, economic, and social consequences of our science, setting a new example by taking responsibility for the larger results of our research, along with credit for the results themselves.

∼

But time is short. Biomedical science is about to change whether or not scientists choose to help steer a new course. The fundamental premise of molecular biology — that any question about a living thing can be answered, any disease understood and eradicated, by learning the detailed interactions of appropriate DNAs, RNAs, and proteins — has grown from an optimistic research strategy to the dominant agenda of a multibillion-dollar research juggernaut. Preventive medicine, public health, and universal education about the workings of the human body have all been overshadowed by the dream that cures for all diseases would flow from an understanding of their molecular mechanisms. The conquests of molecular biology have been many, but in their frothy wake, many implicit promises have gone unmet. Newborn children die in the United States more frequently than they do in more than a dozen other countries, including many that spend far less per person on health care than we do; cancer is not diminishing as a cause of death; and after a decade and billions of dollars, the fact that HIV is more clearly understood as a molecular entity than any other animal virus is small solace to the million Americans who are infected with it or the many millions more who have lost a friend or relative to AIDS.

Molecular biologists have two choices. They can continue working out the meanings of individual genes, going into business for themselves with the discoveries they have made in tax-supported laboratories, shortcutting peer review using news conferences and patent secrecy, taking tax money from their forty million fellow Americans who cannot afford any health insurance at all, and ignoring any potential eugenic uses of their work. Or they can begin to rethink what it is they do and why they do it. An agenda for the next stage of molecular biology is going to be assembled before this century is out, in any event, because the various groups who can pay for basic research — government agencies, agribusiness and pharmaceutical firms, the military, the criminal justice apparatus, would-be eugenicists — need one in order to make their own plans.

These groups will set the future of molecular biology to suit their purposes unless the scientists who would find themselves constrained to carry it out choose to join the debate. The dream of pure and unfettered research is ultimately dangerous; despite the hopes of some of its brightest practitioners, the path of a science — or at least of the science I know — is not self-correcting. Only by openly sharing concerns with others and asking for the broadest possible audience to help set its agenda can science meet its responsibilities to the society that supports it.

≈

I am not sophisticated about music, but when I find something that speaks to me, I listen to it over and over again. Once, as this book was reaching its final state, I heard in Beethoven's Triple Concerto an argument I had been struggling to formulate in words. Beginning with the few genes that are transcribed in the very first cells of an embryo, our bodies play their genomes as if they were instruments playing a concerto or a symphony. Like the Triple Concerto, these melodies of transcription and translation start quietly and simply, then build, interweave, and modulate one another until with a great thumping there is just silence. It follows that since few scientists are Beethovens, the rest of us had best be careful not to bring discord into the wonderfully rich scores that play in each of us every moment of our lives.

The fear of fateful discordances runs deep in our literature. In Part Two of *Faust*, Goethe shows the devil Mephistopheles as a creative if difficult scientist whose research includes a line of experimentation with in vitro homunculi that should by now be startlingly familiar: "Much have I seen, in wandering to and fro, Including crystallized humanity." After being led through many fine experiments by Mephistopheles, Dr. Faust dies a blind fool, thinking he is directing his technicians to produce new crops when in fact he is among zombies who dig his grave. But even as he dies, Faust is busy imagining the culmination of his research plans, sure that if he could make them work, his reputation as a scientist would be secure.

On the other hand, Goethe himself, in conversations recorded by his friend Eckermann, tells us the importance and value of our inability to completely understand his *Faust* and, by extension, why we should not expect to completely understand our genomes: "In such compositions what really matters is that the single masses should be clear and significant, while the whole remains incommensurable; and by that very reason, like an unsolved problem, draws mankind to study it again and again." The issue we face today is whether molecular biology will accept the insights of Goethe or continue to imitate his creations, Mephistopheles and Dr. Faust.

Dr. Faust was only one of a long line of talented but dangerous scientists brought to life in fiction over the years. Many think of Mary Shelley's Dr. Frankenstein — who assembled the parts of corpses into a failed human — as the worst of them and consider his monster the ultimate vision of biomedical science gone awry. But H. G. Wells, a scientifically well-informed writer of the next generation, created in *The Island of Dr. Moreau* a far more chilling and believable vision of the molecular biologist at home in a horrible world of his own making.

Isolated on his island, Dr. Moreau has no wish to touch another person, dead or alive. Instead, he cuts and tweaks the bodies and minds of beasts until they become semihuman. They worship and fear him as their Creator; he modestly accepts this as fitting. But in his experiments on transspecies tissue differentiation the animal sometimes prevails, and once it does, the reverted beasts destroy the doctor, his laboratory, and all else but Prendick, the surviving witness left to tell the tale. Here is how Dr. Moreau explains himself to Prendick:

> "You see, I went on with this research just the way it led me. That is the only way I ever heard of research going. I asked a question, devised some method of getting an answer, and got — a fresh question. Was this possible, or that possible? You cannot imagine what this means to an investigator, what an intellectual passion grows upon him. You cannot imagine the strange colorless delight of these intellectual desires. The thing before you is no longer an animal, a fellow creature, but a problem . . ."
>
> "But," said I, "the thing is an abomination — "
>
> "To this day I have never troubled about the ethics of the matter. The study of Nature makes man at last as remorseless as Nature . . . Each time I dip a living creature into the bath of burning pain, I say, This time I will burn out all the animal, this time I will make a rational creature of my own. After all, what is ten years? Man has been a hundred thousand in the making."

Beyond H. G. Wells and Dr. Moreau, there is of course Aldous Huxley, whose *Brave New World* casts such long shadows over the future of molecular biology. Huxley revisited that book's terrain twenty years after it appeared — soon after the structure of DNA was discovered — and came to this conclusion:

An education for freedom (and for love and intelligence which are at once the conditions and the results of freedom) must be, among other things, an education in the proper uses of language. . . . [This education should teach] the value, first of all, of individual freedom, based on the facts of human diversity and genetic uniqueness; the value of charity and compassion, based on the old familiar fact, lately rediscovered by psychiatry, the fact that, whatever their mental and physical diversity, love is as necessary to human beings as food or shelter; and finally, the value of intelligence, without which love is impotent and freedom unattainable.

Perhaps the first task of the education that Huxley describes is to teach all of us that to be born mortal with a mind that can imagine perfection or immortality is a cruel joke of nature. My colleagues and I can outwit this devil of our imagination, not by serving it, but by understanding that despite their imperfections our children — the world's children — are the only immortality we are allowed. As characters in the literature of DNA, we must work to preserve their futures while we can.

FURTHER READING

The following four lists of references should make it easy to learn more about the subjects and issues of this book. The first is a selection of general textbooks and reviews; the second a collection of books I found particularly helpful while writing this book; the third, four lists covering the main topics of this book; and the last, a collection of important, primary references from scientific journals arranged by chapter.

GENERAL REFERENCES

After fifteen years, the best book for the general reader on the history of molecular biology is still Judson's *Eighth Day of Creation*. For a closer and more recent — but dryer — look at the subject, I recommend Darnell's hefty textbook, *Molecular Cell Biology*. Revised every few years, this single volume captures the excitement of a fast-moving science through its lucid descriptions and colorful illustrations.

The other biology textbooks — Alberts, de Duve, Holtzman, Miklos, and Watson — are each somewhat more specialized than Darnell, but all are easy to find and worth the effort to read. While all biology texts discuss evolution, it is the central theme in Gould, King, Lewontin, Strickberger, and Tattersall. Likewise, Kandel and Gregory are good sources of detailed science concerning the mind and brain, and Miller is an excellent source of information on language. The beautiful pictures in Pauling and Morrison nicely complement the sometimes dense prose of any textbook.

Alberts, B., et al., 1990. *Molecular biology of the cell*, 2d ed. New York: Garland.

Darnell, J., et al., 1990. *Molecular cell biology*, 2d ed. New York: W. H. Freeman.

de Duve, C., 1984. *A guided tour of the living cell*. San Francisco: Freeman.

Gould, S. J., 1977. *Ontogeny and phylogeny*. Cambridge: Belknap Press of Harvard University Press.

Gregory, R. L., ed., 1987. *The Oxford companion to the mind*. New York: Oxford University Press.

Holtzman, E., and A. Novikoff, 1984. *Cells & organelles*, 3d ed. Philadelphia: Saunders.

Judson, H. F., 1979. *The eighth day of creation*. New York: Simon and Schuster.

Kandel, E., and J. Schwartz, 1981. *Principles of neural science*. New York: Elsevier.

King, J. C., 1981. *The biology of race*. Los Angeles: University of California Press.

Lewontin, R., 1982. *Human diversity*. New York: Scientific American Books, W. H. Freeman.

Miklos, D., and G. Freyer, 1990. *DNA science: A first course in recombinant DNA technology*. Cold Spring Harbor, N.Y.: Cold Spring Harbor Laboratory Press.

Miller, G., 1991. *The science of words*. New York: Freeman–Scientific American.

Morrison, P., P. Morrison, and the Office of C. Eames, 1982. *Powers of ten*. San Francisco: Scientific American Press of W. H. Freeman.

Pauling, L., and R. Hayward, 1964. *The architecture of molecules*. San Francisco: W. H. Freeman.

Strickberger, M., 1989. *Evolution*. Boston: Jones and Bartlett.

Tattersall, I., et al., 1986. *Encyclopedia of human evolution and prehistory*. New York: Garland.

Watson, J. D., et al., 1987. *Molecular biology of the gene*. Menlo Park, Calif.: Benjamin/Cummings.

BOOKS TO READ

These are the books from which I have drawn some of my arguments. Many — Dante, Darwin, Gamow, Goethe, Huxley, Koestler, Kuhn, Orwell, Solzhenitsyn, Schrödinger, Watson, Wells — are classics, either of the general culture or of science. The more recent remainder — Calvino, Chatwin, Crick, Edelman, Gore, Gould, Jacob, Rosenfield, Thomas, and Wilson — are likely to be of equally lasting importance.

Calvino, I., 1988. *Six memos for the next millennium.* Cambridge: Harvard University Press.

Chatwin, B., 1988. *The songlines.* New York: Penguin.

Crick, F. R. C., 1988. *What mad pursuit.* New York: Basic Books.

Dante Alighieri, 1954. *The inferno.* Translated by J. Ciardi, reprinted 1982. New York: Signet.

Darwin, Charles, 1859. *The origin of species.* London: John Murray. Reprint, New York: Penguin Classics, 1984.

Edelman, G., 1992. *Bright air, brilliant fire.* New York: Basic Books.

Gamow, G., 1947. *One, two, three . . . infinity.* New York: Viking Press.

Goethe, J., 1975. *Faust.* Translated by Philip Wayne. London: Penguin.

Gore, A., 1992. *Earth in the balance.* Boston: Houghton Mifflin.

Gould, S. J., 1981. *The mismeasure of man.* New York: Norton.

Huxley, A., 1932. *Brave new world.* Reprinted 1989. New York: Harper & Row.

———, 1958. *Brave new world revisited.* Reprinted 1989. New York: Harper & Row.

Jacob, F., 1982. *The logic of life.* New York: Random House.

Koestler, A., 1941. *Darkness at noon.* New York: Macmillan.

Kuhn, T., 1962. *The structure of scientific revolutions.* International Encyclopedia of Unified Science 2, reprinted 1970. Chicago: University of Chicago Press.

Orwell, G., 1949. *1984.* London: Martin Secker and Warburg.

Rosenfield, I., 1992. *The strange, familiar and forgotten*. New York: Knopf.

Schrödinger, E., 1944. *What is life?* Cambridge, Eng.: Cambridge University Press. Reprint, Cambridge, Eng.: Cambridge University Press, 1988.

Snow, C. P., 1959. *The two cultures and the scientific revolution*. Cambridge, Eng.: Cambridge University Press.

Solzhenitsyn, A., 1968. *The first circle*. New York: Harper & Row.

Thomas, L., 1974. *Lives of a cell*. New York: Viking.

———, 1983. *The youngest science*. New York: Viking.

Watson, J. D., 1968. *The double helix*. New York: Atheneum. Reprint, New York: Norton Critical Edition, ed. G. Stent, 1980.

Wells, H. G., 1984 (reprint). *The island of Dr. Moreau*. New York: Signet.

Wilson, E. O., 1992. *The diversity of life*. Cambridge: Harvard University Press.

SPECIALIZED BACKGROUND BOOKS

These four sets of books each delve into specifics of one of the four major topics of this book: molecular biology and genetics, evolution and the history of life on earth, the nature and history of language, and the politics of medicine and science. The politics — or, to the fastidious, the ethics — of medical science is a booming subject in its own right; by far the greatest number of books in these lists are ones that attempt to map the boundaries of this branch of science.

Molecular Biology and Genetics

Alberts, B., ed., 1988. *Mapping and sequencing the human genome*. Washington, D.C.: National Research Council, National Academy of Sciences.

Bonner, J. T., 1974. *On development*. Cambridge: Harvard University Press.

Brandon, L., 1991. *Introduction to protein structure*. New York: Garland Press.

Buchsbaum, R., 1948. *Animals without backbones.* Chicago: University of Chicago Press.

de Pomerai, D., 1985. *From gene to animal.* Cambridge, Eng.: Cambridge University Press.

Gierasch, L., and J. King, eds., 1990. *Protein folding.* Washington, D.C.: AAAS.

Keller, E., 1983. *A feeling for the organism.* New York: W. H. Freeman.

Krimsky, S., 1983. *Genetic alchemy.* Cambridge: MIT Press.

Levine, A. J., ed., 1992. *Tumor suppressor genes, the cell cycle and cancer.* Cold Spring Harbor, N.Y.: Cold Spring Harbor Laboratory Press.

National Research Council, 1992. *DNA technology in forensic science.* National Academy of Sciences Press.

Office of Technology Assessment, 1988. *Mapping our genes: Genome projects, how big, how fast?* Baltimore: Johns Hopkins University Press.

Office of Technology Assessment, 1992. *A new technological era for American agriculture.* Washington, D.C.: U.S. Congress.

Office of Technology Assessment, 1992. *Cystic fibrosis and DNA tests: Implications of carrier screening.* Washington, D.C.: U.S. Congress.

Pollack, R., 1981. *Readings in mammalian cell culture,* 2d ed. Cold Spring Harbor, N.Y.: Cold Spring Harbor Press.

Tooze, J., ed., 1980. *Molecular biology of tumor viruses.* Cold Spring Harbor, N.Y.: Cold Spring Harbor Press.

Evolution and the History of the Earth

Albritton, C. C., Jr., 1980. *The abyss of time.* San Francisco: Freeman, Cooper. Reprint, Los Angeles: Tarcher, 1986.

Clapham, W., 1981. *Human ecosystems.* New York: Macmillan.

Dawkins, R., 1976. *The selfish gene.* New York: Oxford.

Ereshevsky, M., 1992. *The units of evolution: Essays on the nature of species.* Cambridge: MIT Press.

Lewontin, R., 1974. *The genetic basis of evolutionary change.* New York: Columbia University Press.

Li, W.-H., and D. Graur, 1991. *Fundamentals of molecular evolution.* Sunderland, Mass.: Sinauer.

Margulis, L., and K. Schwartz, 1982. *Five kingdoms, An illustrated guide to the phyla of life on Earth.* New York: W. H. Freeman.

————, and L. Olendzenski, eds., 1992. *Environmental evolution.* Cambridge: MIT Press.

Mayr, E., 1988. *Toward a new philosophy of biology.* Cambridge: Harvard University Press.

Selander, R., A. Clark, and T. Whittam, 1991. *Evolution at the molecular level.* Sunderland, Mass.: Sinauer.

Origins and Mechanisms of Language

Bruner, J., 1990. *Acts of meaning.* Cambridge: Harvard University Press.

Delbrück, M., 1975. *Mind from matter?* Palo Alto: Blackwell.

Johnson, G., 1992. *In the palaces of memory.* New York: Vintage.

Klima, E., and U. Bellugi, 1978. *The signs of language.* Cambridge: Harvard University Press.

Landau, M., 1991. *Narratives of evolution.* New Haven: Yale University Press.

Politics of Science and Medicine

Angier, N., 1988. *Natural obsessions.* Boston: Houghton Mifflin.

Annas, G., 1992. *Gene mapping: Using law and ethics as guides.* New York: Oxford University Press.

Bankowski, Z., and A. Capron, eds., 1991. *Genetics, ethics and human values: Human genome mapping, genetic screening and gene therapy.* World Health Organization.

Bulger, R., E. Heitman, and S. J. Reiser, 1993. *The ethical dimensions of the biological sciences.* New York: Cambridge University Press.

Davis, B., ed., 1992. *The genetic revolution: Scientific prospects and public perception.* Baltimore: Johns Hopkins University Press.

de Solla Price, D. J., 1963. *Little science, big science.* New York: Columbia University Press.

Dubos, R., 1961. *The dreams of reason.* New York: Columbia University Press.

Ginzberg, E., 1990. *The medical triangle.* Cambridge: Harvard University Press.

Hall, S., 1992. *Mapping the next millennium: The discovery of the new geographics.* New York: Random House.

Holtzman, N., 1989. *Proceed with caution: Predicting genetic risks in the recombinant DNA era.* Baltimore: Johns Hopkins.

Kevles, D., and L. Hood, eds., 1992. *The code of codes: Scientific and social issues in the human genome project.* Cambridge: Harvard University Press.

Kevles, D. J., 1985. *In the name of eugenics.* New York: Knopf.

Lange, J., et al., 1940. *Erbpathologie.* Vol. 1 of *Menschliche erblehre und rassenhygiene.* Munich: Lehmanns.

Lappe, M., 1984. *Broken code: The exploitation of DNA.* San Francisco: Sierra Club Books.

Lee, Thomas, 1992. *Human genome project: Quest for the code of life.* New York: Plenum.

Lifton, R., 1986. *The Nazi doctors.* New York: Basic Books.

Müller-Hill, B., 1988. *Murderous science.* New York: Oxford University Press.

Nelkin, D., and L. Tancredi, 1989. *Dangerous diagnostics.* New York: Basic Books.

Nichols, E. K., 1988. *Human gene therapy.* Cambridge: Harvard University Press.

Pollack, R., et al., 1973. *Biohazards in biological research.* Cold Spring Harbor, N.Y.: Cold Spring Harbor Press.

Proctor, R., 1988. *Racial hygiene.* Cambridge: Harvard University Press.

Rainger, R., 1992. *An agenda for antiquity.* University of Alabama Press.

Suzuki, D., and P. Knudtson, 1989. *Genethics: The clash between the new genetics and human value.* Cambridge: Harvard University Press.

Weatherall, D., 1991. *The new genetics and clinical practice.* New York: Oxford University Press.

Wingerson, L., 1990. *Mapping our genes.* New York: Plume.

JOURNAL ARTICLES

These lists would not be complete without a sampling, by chapter, of information written by one scientist for another. I drew many of my examples of DNA as a language from this primary scientific literature. These papers are for the most part also good examples of the style by which — in fits and starts — the molecular biology of our species advances.

Chapter 1

Watson, J. D., and F.H.C. Crick, 1953. A structure for deoxyribose nucleic acid. *Nature*, April 25: 737–38.

———. Genetical implications of the structure of deoxyribonucleic acid. *Nature*, May 30: 964–67.

Chapter 2

Charlesworth, B., 1991. The evolution of sex chromosomes. *Science* 251: 1030–33.

Clark, L., et al., 1992. Defective epithelial chloride transport in a gene-targeted mouse model of cystic fibrosis. *Science* 257: 1125–30.

Diamond, J., 1991. Curse and blessing of the ghetto. *Discover*, March 1991: 60–65.

Gusella, J., et al., 1983. A polymorphic DNA marker genetically linked to Huntington's disease. *Nature* 306: 234–38.

Hamer, D., et al., 1993. A linkage between DNA markers on the X chromosome and male sexual orientation. *Science* 261: 5119–22.

Harley, V., et al., 1992. DNA binding activity of recombinant SRY from normal males and XY females. *Science* 255: 453–56.

Merz, B., 1987. Matchmaking scheme solves Tay-Sachs problem. *J. Amer. Med. Assoc.* 258: 2636–37.

News item, 1968. Logic of biology. *Nature* 220: 429–30.

Potter, H., 1991. Review and hypothesis: Alzheimer disease and Down syndrome: Chromosome 21 nondisjunction may underlie both disorders. *Am. J. Human Genetics* 48: 1192–1200.

Rommens, J. M., et al., 1989. Identification of the cystic fibrosis gene: Chromosome walking and jumping. *Science* 245: 1059–80.

Chapter 3

Aggarwal, A., et al., 1988. Recognition of a DNA operator by the repressor of phage 434: A view at high resolution. *Science* 242: 899–906.

Benner, M., et al., 1989. Modern metabolism as a palimpsest of the RNA world. *Proc. Nat. Acad. Sci.* 86: 7054.

Brendel, V., et al., 1986. Linguistics of nucleotide sequences: Morphology and comparison of vocabularies. *J. Biomolecular Structure and Dynamics* 4: 11–21.

Darnell, J., 1982. Variety in the level of gene control in eukaryotic cells. *Nature* 297: 499–506.

De Duve, C., et al., 1988. The second genetic code. *Nature* 333: 117–18.

Dombroski, B., et al., 1991. Isolation of an active human transposable element. *Science* 254: 1805–7.

Ealick, S., et al., 1991. Three-dimensional structure of recombinant human interferon-γ. *Science* 252: 698–702.

Francklyn, E., et al., 1992. Overlapping nucleotide determinants for specific aminoacylation of RNA microhelices. *Science* 255: 1121–25.

Fu, Y.-H., et al., 1992. An unstable triplet repeat in a gene related to myotonic muscular dystrophy. *Science* 255: 1256–58.

Gething, M., and J. Sambrook, 1992. Protein folding in the cell. *Nature* 355: 33–45.

Gogos, J., et al., 1992. Sequence discrimination by alternatively spliced isoforms of a DNA binding zinc finger domain. *Science* 257: 1951–55.

Hanscombe, O., et al., 1991. Importance of globin gene order for correct developmental expression. *Genes & Development* 5: 1387–94.

Hard, T., et al., 1990. Solution structure of the glucocorticoid receptor DNA-binding domain. *Science* 249: 157–60.

Helfman, D., et al., 1988. Alternative splicing of tropomyosin pre-mRNAs in vitro and in vivo. *Genes & Development* 2: 1627–38.

Jurka, J., 1990. Novel families of interspersed repetitive elements from the human genome. *Nucleic Acids Research* 18: 137–41.

Koch, C., et al., 1991. SH2 and SH3 domains: Elements that control interactions of cytoplasmic signalling proteins. *Science* 252: 668–74.

Kuhl, P., et al., 1992. Linguistic experience alters phonetic perception in infants by 6 months of age. *Science* 252: 606–8.

McClintock, B., 1984. The significance of responses of the genome to challenge. *Science* 226: 792–801.

Nathans, J., et al., 1989. Molecular genetics of human blue cone monochromacy. *Science* 245: 831–38.

Piatigorsky, J., and G. Wistow, 1991. The recruitment of crystallins: New functions precede gene duplication. *Science* 252: 1078–79.

Ptashne, M., and A. Gann, 1990. Activators and targets. *Nature* 346: 329–31.

Reilly, J., et al., 1990. Once more with feeling: Affect and language in atypical populations. *Development and Psychopathology* 2: 367–91.

Schulman, L., and H. Pelka, 1989. The Anticodon contains a major element of the identity of Arginine Transfer RNAs. *Science* 246: 1595–97.

Shih, M.-C., et al., 1988. Intron existence predated the divergence of eukaryotes and prokaryotes. *Science* 242: 1164–66.

Spector, D., 1990. Higher order nuclear organization: Three dimensional distribution of small nuclear ribonucleoprotein particles. *Proc. Nat. Acad. Sci.* 87: 147–51.

Stanfield, R., et al., 1990. Crystal structures of an antibody to a peptide and its complex with peptide antigen at 2.8 Å. *Science* 248: 712–19.

Thal, D., et al., 1989. Language and cognition in two children with William's syndrome. *J. Speech and Hearing Res.* 32: 489–500.

Thibodeau, S., et al., 1993. Microsatellite instability of cancer of the proximal colon. *Science* 260: 816–22.

Treisman, J., et al., 1989. A single amino acid can determine the DNA binding specificity of homeodomain proteins. *Cell* 59: 553–62.

Vinson, C., et al., 1989. Scissors-grip model for DNA recognition by a family of leucine zipper proteins. *Science* 246: 911–16.

Zahler, A., et al., 1993. Distinct functions of SR proteins in alternative pre-mRNA splicing. *Science* 360: 219–22.

Chapter 4

Blaese, M., 1993. Development of gene therapy for immunodeficiency: Adenosine deaminase deficiency. *Pediatric Research* 33 (suppl.): S49–S55.

Brosius, J., 1991. Retroposons — seeds of evolution. *Science* 251: 753–54.

Cantor, C., 1990. Orchestrating the human genome project. *Science* 248: 49–51.

Chakraborty, R., and K. Kidd, 1991. The utility of DNA typing in forensic work. *Science* 254: 1735–39.

Davis, B., et al., 1990. The human genome and other initiatives. *Science* 342–43.

Engelhardt, J., et al., 1993. Direct gene transfer of human CFTR into human bronchial epithelia of xenografts with E1-deleted adenoviruses. *Nature Genetics* 4: 27–34.

Erlich, H., 1991. Recent advances in the Polymerase Chain Reaction. *Science* 252: 1643–50.

Jeskevich, J., and C. Guyer, 1990. Bovine growth hormone: Human food safety evaluation. *Science* 249: 875–84.

Kieleczawa, J., et al., 1992. DNA sequencing by primer walking with strings of contiguous hexamers. *Science* 258: 1787–91.

Le Gal La Salle, G., et al., 1993. An adenovirus vector for gene transfer into neurons and glia in the brain. *Science* 259: 988–90.

Lewontin, R., and D. Hartl, 1991. Population genetics in forensic DNA typing. *Science* 254: 1745–50.

Li, H., et al., 1988. Amplification and analysis of DNA sequences in single human sperm and diploid cells. *Nature* 335: 414–17.

Lusher, J., et al., 1993. Recombinant factor VIII for the treatment of previously untreated patients with hemophilia A. *New England J. Medicine* 328: 453–59.

NIH/CEPH Collaborative Mapping Group, 1992. A comprehensive genetic linkage map of the human genome. *Science* 258: 67–86; appendix, 148–62.

Olson, M., et al., 1989. A common language for physical mapping of the human genome. *Science* 245: 1434–35.

Patterson, A., et al., 1988. Resolution of quantitative traits into Mendelian factors by using a complete linkage map of restriction fragment length polymorphisms. *Nature* 335: 721.

Radmacher, M., et al., 1992. From molecules to cells: Imaging soft samples with the atomic force microscope. *Science* 257: 1900–05.

Reilly, P., 1992. ASHG statement on genetics and privacy: Testimony to United States Congress. *Am. J. Human Genetics* 50: 640–42.

Risch, N., and B. Devlin, 1992. On the probability of matching DNA fingerprints. *Science* 255: 717–20.

Vasil, I., 1990. The realities and challenges of plant biotechnology. *Bio/technology* 8: 296–301.

Watson, J. D., 1990. The human genome project: Past, present and future. *Science* 248: 44–48.

Weissenbach, J., et al., 1992. A second-generation linkage map of the human genome. *Nature* 359: 794–801.

Chapter 5

Aaltonen, L., et al., 1993. Clues to the pathogenesis of familial colorectal cancer. *Science* 260: 812–16.

Aldous, P., 1993. European biotech: Thumbs down for cattle hormone. *Science* 261: 418.

Anderson, C., 1993. Researchers win decision on knockout mouse pricing. *Science* 260: 23–24.

Behringer, R., et al., 1989. Synthesis of functional human hemoglobin in transgenic mice. *Science* 245: 971–73.

Borowiec, J., et al., 1990. Binding and unwinding: How T antigen engages the SV40 origin of DNA replication. *Cell* 60: 181–84.

Chisaka, O., and M. Capecchi, 1991. Regionally restricted developmental defects resulting from targeted disruption of the mouse homeobox gene *hox-1.5*. *Nature* 350: 473–79.

Daubert, William, et al., petitioners, v. Merrell Dow Pharmaceuticals, Inc. Supreme Court decision 92–102, decided June 28, 1993. *United States Law Week* 6-29-93: 4805–11.

DePamphilis, M., and M. Bradley, 1986. SV40. Chapter 3 in *The papovaviridae*, vol. I. New York: Plenum.

Dyson, N., et al., 1990. Large T antigens of many polyoma viruses are able to form complexes with the retinoblastoma protein. *J. Virol.* 64: 1353–56.

Frebourg, T., et al., 1992. Germ-line mutations of p53 suppressor gene in patients with high risk for cancer inactivate the p53 protein. *Proc. Nat. Acad. Sci.* 89: 6413–17.

Galton, D., 1988. Molecular genetics of coronary heart disease. *Euro. J. Clin. Invest.* 18: 219–25.

Grant, S., et al., 1992. Impaired long-term potentiation, spatial learning, and hippocampal development in fyn mutant mice. *Science* 258: 1903–10.

Handyside, A., et al., 1992. Birth of a normal girl after in vitro fertilization and pre-implantation testing for cystic fibrosis. *New England J. Medicine* 327: 905–9.

Hiatt, A., et al., 1989. Production of antibodies in transgenic plants. *Nature* 342: 76–78.

Jaenisch, R., 1988. Transgenic animals. *Science* 240: 1468–74.

Kawabata, S., et al., 1991. Amyloid plaques, neurofibrillary tangles and neuronal loss in brains of transgenic mice overexpressing a C-terminal fragment of human amyloid precursor protein. *Nature* 354: 476–78.

———, 1992. Alzheimer's retraction. *Nature* 356: 23.

Kessel, M., and P. Gruss, 1990. Murine developmental control genes. *Science* 249: 374–79.

Kuerbitz, S., et al., 1992. Wild-type p53 is a cell cycle check-point determinant following irradiation. *Proc. Nat. Acad. Sci.* 89: 7491–95.

Leonard, J. M., et al., 1988. Development of disease and virus recovery in transgenic mice containing HIV proviral DNA. *Science* 242: 1665–70.

Levine, A., et al., 1991. The p53 tumour suppressor gene. *Nature* 351: 453–56.

Loeber, G., et al., 1989. The zinc finger region of SV40 large T antigen. *J. Virology* 63: 94–100.

Mason, H., et al., 1992. Expression of hepatitis B surface antigen in transgenic plants. *Proc. Nat. Acad. Sci.* 89: 11745–49.

McVey, D., et al., 1989. Properties of the DNA-binding domain of the SV40 large T antigen. *Mol. Cell. Biol.* 9: 5525–36.

Padgette, S., et al., 1989. Selective herbicide tolerance through protein engineering. *Cell Cult. Somatic Cell Gen. Plants* 6: 441–76.

Paszkowski, J., et al., 1988. Gene targeting in plants. *EMBO J.* 7: 4021–26.

Peltomaki, P., et al., 1993. Genetic mapping of a locus predisposing to human colorectal cancer. *Science* 260: 810–12.

Qian, Y., et al., 1989. The structure of the Antennapedia homeo domain determined by NMR spectroscopy: Comparison with prokaryotic repressors. *Cell* 59: 573–80.

Ratcliff, R., et al., 1993. Production of severe cystic fibrosis mutation in mice by gene targeting. *Nature Genetics* 4: 35–41.

Sepulveda, A., et al., 1989. Development of a transgenic mouse system for the analysis of stages of liver carcinogenesis using tissue-specific expression of SV40 Large T-antigen controlled by regulatory elements of the human α-1-antitrypsin gene. *Cancer Research* 49: 6108–17.

Vogelstein, B., and K. Kinzler, 1992. p53 function and dysfunction. *Cell* 70: 523–26.

Walsh, C., and C. Cepko, 1992. Widespread dispersion of neuronal clones across functional regions of the cerebral cortex. *Science* 255: 434–40.

Yokoyama, T., et al., 1993. Reversal of left-right asymmetry: A sinus inversus mutation. *Science* 260: 679–82.

Chapter 6

Abelson, J., 1990. Directed evolution of nucleic acids by independent replication and selection. *Science* 249: 488–89.

Aldhous, P., 1992. The promise and pitfalls of molecular genetics. *Science* 257: 164–65.

Asfaw, B., et al., 1992. The earliest Acheulean from Konso-Gardula. *Nature* 330: 732–35.

Ayala, F., 1986. On the virtues and pitfalls of the molecular evolutionary clock. *J. Heredity* 77: 226–35.

Bailey, J., et al., 1993. Heritable factors influence sexual orientation in women. *Arch. Gen. Psychiatry* 50: 217–23.

Barbujani, G., and R. Sokal, 1990. Zones of sharp genetic change in Europe are also linguistic boundaries. *Proc. Nat. Acad. Sci.* 87: 1816–19.

Begun, D., 1992. Miocene fossil hominids and the chimp-human clade. *Science* 257: 1929–32.

Bouchard, T., et al., 1990. Sources of human psychological differences: The Minnesota study of twins reared apart. *Science* 250: 223–28.

Caldeira, K., et al., 1992. The life span of the biosphere revisited. *Nature* 360: 721–23.

Cann, R., et al., 1987. Mitochondrial DNA and human evolution. *Nature* 325:31–36.

Cavalli-Sforza, L., 1990. How can one study individual variation for 3 billion nucleotides of the human genome? *Am. J. Human Genetics* 46: 649–51.

———, et al., 1988. Reconstruction of human evolution: Bringing together genetic, archaeological and linguistic data. *Proc. Nat. Acad. Sci.* 85: 6002–6.

———, 1993. Demic expansions and human evolution. *Science* 259: 639–46.

Churchland, P., and T. Sejnowski, 1988. Perspectives on cognitive neuroscience. *Science* 242: 741–45.

Corina, D., et al., 1992. The linguistic basis of left hemisphere specialization. *Science* 255: 1258–60.

Coyne, J., 1992. Genetics and speciation. *Nature* 355: 511–15.

Deacon, T., 1990. Rethinking mammalian brain evolution. *Amer. Zool.* 30: 629–705.

DeSalle, R., et al., 1992. DNA sequences from a fossil termite in Oligo-Miocene amber and their phylogenetic implications. *Science* 257: 1933–37.

Gazzaniga, M., 1989. The organization of the human brain. *Science* 245: 947–52.

Gingerich, P., 1985. Species in the fossil record: Concepts, trends and transitions. *Paleobiology* 11: 27–41.

Goodman, M., 1989. Emerging alliance of phylogenetic systematics and molecular biology: A new age of exploration. *The hierarchy of life,* eds. B. Fernholm et al., ch. 4: 43–61. New York: Elsevier.

Green, P., et al., 1993. Ancient conserved regions in new gene sequences and the protein databases. *Science* 259: 1711–16.

Hall, S., 1992. How technique is changing science. *Science* 257: 344–49.

Han, T.-M., et al., 1992. Megascopic eukaryotic algae from the 2.1-billion-year-old Negaunee iron-formation, Michigan. *Science* 257: 232–35.

Handyside, A., et al., 1989. Biopsy of human preimplantation embryos and sexing by DNA amplification. *Lancet* Feb. 18: 347–49.

Knoll, A., 1992. The early evolution of eukaryotes: A geological perspective. *Science* 256: 622–27.

Lake, J., 1990. Origin of the metazoa. *Proc. Nat. Acad. Sci.* 87: 763–66.

Lander, E., and D. Botstein, 1989. Mapping mendelian factors underlying quantitative traits using RFLP linkage maps. *Genetics* 121: 185.

Lightman, A., and O. Gingerich, 1991. When do anomalies begin? *Science* 255: 690–95.

May, R., 1988. How many species are there on Earth? *Science* 241: 1441–49.

Morton, N., et al., 1993. Kinship bioassay on hypervariable loci in Blacks and Caucasians. *Proc. Nat. Acad. Sci.* 90: 1892–96.

Murtha, M., et al., 1991. Detection of homeobox genes in development and evolution. *Proc. Nat. Acad. Sci.* 88: 10711–15.

Novacek, M., 1992. Mammalian phylogeny: Shaking the tree. *Nature* 356: 121–25.

Pimm, S., and J. Gittleman, 1992. Biological diversity: Where is it? *Science* 255: 940.

Pinker, S., and P. Bloom, 1990. Natural language and natural selection. *Behav. and Brain Sci.* 13: 707–84.

Plomin, R., 1990. The role of inheritance in behavior. *Science* 248: 183–88.

Poinar, G., et al., 1993. Terrestrial soft-bodied protists and other microorganisms in Triassic amber. *Science* 259: 222–24.

Risch, N., 1992. Genetic linkage: Interpreting Lod scores. *Science* 255: 803–4.

Schopf, W., 1993. Microfossils of the early Archean Apex Chert: New evidence for the antiquity of life. *Science* 260: 640–46.

Sleep, N., et al., 1989. Annihilation of ecosystems by large asteroid impacts on the early Earth. *Nature* 342: 139–42.

Smouse, P., et al., 1982. Multiple-locus allocation of individuals to groups as a function of the genetic variation within and differences among human populations. *Amer. Nat.* 119: 445–63.

Sokal, R., et al., 1990. Genetics and language in European populations. *Amer. Nat.* 135: 157–75.

Tianyuan, L., and D. Etler, 1992. New middle Pleistocene hominid crania from Yunxian in China. *Nature* 357: 404–7.

van den Bergh, S., 1992. The age and size of the universe. *Science* 258: 421–23.

Vigilant, L., et al., 1991. African populations and the evolution of human mitochondrial DNA. *Science* 253: 1503–7.

Wood, B. 1992. Origin and evolution of the genus *Homo*. *Nature* 355: 783–90.

Yoon, C., 1993. Counting creatures great and small. *Science* 260: 620–22.

Conclusion

Brown, G., 1992. Rational science, irrational reality: A congressional perspective on basic research and society. *Science* 258: 200–01.

Glass, B., 1971. Science: Endless horizons or golden age? *Science* 171: 23–29.

Lederman, L., 1992. The advancement of science. *Science* 256: 1119–24.

INDEX